Lewis Wright

The Induction Coil in Practical Work

Including Röntgen X-Rays

Lewis Wright

The Induction Coil in Practical Work
Including Röntgen X-Rays

ISBN/EAN: 9783744744294

Printed in Europe, USA, Canada, Australia, Japan

Cover: Foto ©berggeist007 / pixelio.de

More available books at **www.hansebooks.com**

THE INDUCTION COIL IN PRACTICAL WORK

INCLUDING RÖNTGEN X RAYS

BY

LEWIS WRIGHT

AUTHOR OF "LIGHT: A COURSE OF EXPERIMENTAL OPTICS"
"OPTICAL PROJECTION"
"A POPULAR HANDBOOK TO THE MICROSCOPE"
ETC. ETC.

London
MACMILLAN AND CO. Limited
NEW YORK: THE MACMILLAN COMPANY
1897

All rights reserved

PREFACE

It would not have occurred to myself to prepare this unpretentious little handbook. It was proposed to me, with the idea and request that it should be simple and practical, and compressed within specified limits.

These conditions define its character and object. It is not even an elementary treatise upon electricity, or the theory and construction of transformers, or the Röntgen rays. It is written simply and solely as a practical help to the efficient and safe use of an Induction Coil, with some special reference to the revived and extensive use of that apparatus in surgical and physiological work with Röntgen rays. This new field of experiment has brought many into personal contact with coils, who have never had any acquaintance with such instruments before. Not a few of such have actually stated their need of such information as it is here attempted to supply; and it is thought that some will like to have a convenient outline of the many other impressive and beautiful experiments, in which the Induction Coil bears a part.

The last chapter has occasioned most anxiety, though

circumstances have enabled me to follow the gradual advances in Radiography almost from the first, and to keep up some acquaintance with the voluminous literature upon the subject. All who have done so, however, know what a mass of the latter has been superseded by more perfect knowledge, and it would be vain to hope that in all cases I have judged correctly as to what was best worth recording. But I have done my best to separate the wheat from the chaff, and embody in small space and practical hints what may be most useful to the actual worker. For whatever error in judgment, or ultimately of fact, may be discovered, I can only plead my own share of the ignorance which still exists on many points, and do my best to amend it as opportunity may be afforded, and exact knowledge of the subject shall extend.

L. W.

LONDON, *April* 10, 1897.

CONTENTS

CHAPTER
ELECTRICAL INDUCTION 1

CHAPTER II
INDUCTION COILS 23

CHAPTER III
CHOICE, MANIPULATION, AND CARE OF COILS 46

CHAPTER IV
MISCELLANEOUS EXPERIMENTS 57

CHAPTER V
THE DISCHARGE IN PARTIAL VACUA 70

CHAPTER VI
SPECTRUM WORK 88

CHAPTER VII
THE DISCHARGE IN HIGH VACUA 94

CHAPTER VIII
RÖNTGEN X RAYS 125

PLATES

PLATE I.—HAND . . . . *To face page* 125

,, II.—FINGERS, SHOWING STRUCTURE OF BONES ,, 149

,, III.—SMALL SNAKE . . . . ,, 152

,, IV.—PELVIC REGION OF BOY . . . . *Frontispiece*

THE INDUCTION COIL IN PRACTICAL WORK

CHAPTER I

ELECTRICAL INDUCTION

To use an Induction Coil safely and intelligently, something must be known not only of its construction, but of the principles applied and involved in this. We may learn sufficient of these for practical working purposes without going beyond very broad and popular outlines.

1. **Electricity.**—No one knows to this day what is the absolute reality we glibly speak of as "Electricity." At the celebration of his professorial jubilee, in 1896, Lord Kelvin (who knows as much about it as any man) said frankly, "I know no more of electric and magnetic force, nor of the relation between ether, electricity, and ponderable matter, nor of chemical affinity, than I knew and tried to teach my students fifty years ago, in my first session as professor." But we do know a great deal more *about* Electricity in detail; and what we have learnt has caused a general trend of opinion which manifests a curious return towards older ideas, and which in several respects makes easier some general understanding of the subject by the ordinary reader. In the text-books of fifty years ago we found a material theory of one

or two "fluids," as postulated by Du Fay, Symmer, and Franklin; the excess or defect of one fluid, or the complementary proportions of two, interpreting the phenomena. This was given up for the teaching that Electricity was "a form of Energy," a doctrine which held the field for many years. This must perforce be given up in its turn, for views which in some respects more resemble those of the older philosophers.

We may indeed be said to know that Electricity is *not* a form of Energy. It never behaves as such, but always as an Entity of some kind. Any form of Energy can be created or destroyed, by some form of Work transforming it to or from some other form. But Electricity is never created or destroyed. Its stresses or its motions—what we call "*electrification*"—are forms of Energy, but so are stresses or motions in Matter or Ether. Its motion meets with resistance and causes friction, and heat like other friction; and some phenomena (such as "extra current" and oscillatory discharge) appear to show that it possesses (at all events in connection with Matter or Ether) "inertia" or momentum. These of course are forms of Energy, which appear and disappear; but Electricity itself cannot be destroyed.

Hence we have to regard it as an actual entity of some kind. As Dr. Lodge says,[1] it may possibly be some imponderable form of Matter; or it may possibly be, as some thing, the Ether itself; or it may be (and perhaps most probably is) some persistent Manifestation of the Ether to us. We have, any way, to think of it as something in which opposite or complementary strains and stresses and motions can be produced, and whose *movements or stresses* are forms of Energy, and produce physical effects.

2. **Dual Constitution.**—The phenomena also compel belief in some inscrutable, but essential and absolute *doubleness* of constitution, such that setting up a stress of one kind,

[1] *Modern Views of Electricity*. (Macmillan.)

absolutely involves the setting up of an equal stress somewhere, of an exactly opposite kind. Not something exactly *similar* (except in amount), but in some way *opposite*, so that the two together are neutralised; the "two-fluid" theory is impossible. Many *different* phenomena in positive and negative charges and discharges (such as dust figures, brush and glow in vacuum tubes, unequal rate of dissipation, unequal action of heat or Röntgen rays upon them, &c.) shut us up to some kind of *positive* and *negative*, rather than two equal similar affections.

Supposing Electrification to be the setting up of a stress in the Ether, necessarily linked to an opposite stress, we may conceive that stress as (*a*) maintained stationary; (*b*) translated through space by the help of some medium; (*c*) moved in a circle or rotated; (*d*) rapidly alternated or vibrated, like the stresses in a vibrating spring. This conception of Dr. Lodge is perhaps as near as we can get at present to the probable reality, and would represent—

(*a*) Fixed stress . . . . Static electricity or charge.
(*b*) Translatory motion . Electric current.
(*c*) Rotation . . . . . Magnetism.
(*d*) Vibration . . . . . Radiation (including light, heat, Röntgen, and electric rays).

3. **Charge, Discharge, and Current.**—It is well to know this much of the general way in which physicists are coming to regard Electricity. But our brief practical pages have chiefly to do with Electric Current, and for plain working purposes a rougher, and not strictly accurate, but simpler and symbolic idea, will be found sufficient. As Franklin saw, Electricity really *behaves* in many respects[1] like an incompressible liquid, standing normally at one uniform level. Consider this confined in separate tanks, or divided in portions by partitions (as electricity appears to our senses "bound" by

[1] Only partially. We cannot conceive of two "opposite" fluids; and no fluid repels itself.

ponderable matter). Then we can only raise the level in one portion, by taking an equal quantity out of another portion, and between the two is now a difference of levels. That process has needed Work, and the Energy is now stored in this difference of levels, which causes pressure or stress. The high level presses downwards and the low level upwards. This may represent an electric "charge."

Such must be accompanied by a tendency or effort to *discharge*, or return to the normal level from both levels. We may conceive the partition *bursting*, when the higher level will rapidly sink and the other rise, but the motion will not instantaneously stop. The low portion of the liquid will rise too high and the higher sink too low, by the *momentum* created; and this will go on through many oscillations before all comes to rest. Such will represent a sudden discharge of static electricity, as of a Leyden jar, which can be shown to have such an oscillatory character by means of a rapidly-revolving mirror.

But level may be restored in another way; by connecting the two portions through pipes, or holes in the partition. Then we shall have a *flow* of the fluid, which will meet with resistance in the pipe, causing friction and heat. If we further conceive of some form of work, such as that of a pump, keeping up the higher level from the lower as fast as it is lowered by the flow, we shall have a steady *current* through the pipe. There is really no essential difference between the sudden fall and a current. Fall is a very rapid flow, while current is a slower because limited flow. It is so with static "discharge," and "current" in electricity. The more essential analogies are those following.

4. **Measurement of Currents and Conductors.**—Difference of levels causes pressure or stress, which is proportionate to the difference. This *pressure* in electricity is called potential, or potential difference (expressed as P.D.); and the force due to it, electro-motive force (expressed as

E.M.F.). It was formerly called "intensity" of current, but this word has now another meaning. Such "pressure" can be measured, as we measure that of the air. In our currents it is measured as so many *volts*.

We may consider along with this measure the *resistance* due to friction between the liquid and the pipe. A certain quantity per second can only be driven through a given pipe by a certain pressure, which has to overcome the resistance as well as cause a flow. In an electric conductor resistance is measured in *ohms*.

Finally we have to consider the *quantity* of liquid flowing, either absolutely or per second, through the pipe; and it will be manifest that the same quantity may flow in a given time through a large pipe with a low pressure, as through a smaller pipe if more pressure causes more rapid flow. The electrical unit of absolute quantity is called a *coulomb*, and the quantity flowing per second, or any other period of time, is measured in *amperes*.

All these measures are related to each other. The *coulomb* is the quantity of current which, when employed in the electrolysis of a 25 per cent. solution of silver nitrate, will deposit 0·00118 of a gramme of metallic silver. An *ampere* is the quantity and rate of current which will do this amount of work every second. The *volt* is a pressure or E.M.F. or P.D., about equal to a cell of Daniell's battery, or half that of a Grove or Bunsen battery (§ 28). One *ohm* is the resistance to a current offered by a column of mercury 1 mm. square in section, and 106·3 cm. in length, at freezing point. These concrete measures are given to help in fixing definite ideas. It need only be added that one *volt* and one *ohm* are related to the other measures by the definition, that 1 volt is the E.M.F. which maintains an effective current of 1 ampere through a conductor offering a resistance of 1 ohm. We need not discuss the other relations which grow out of these.

It will be clear that the same amount of work may be

obtained by using 10 ampères at 5 volts, as by 5 amperes at 10 volts; just as we get the same work from a small quantity of water with a high fall, as from a larger quantity with little fall. The object of the Induction Coil to be dealt with in these pages, is to transform currents of large quantity at low pressure into much smaller currents at high pressure. Other coils, usually then called " transformers," may do precisely the opposite; the reasons for transformation being in either case that while the resulting amount of work when transformed will be the same (less some amount always dissipated when we transform Energy), the higher pressure or the larger quantity are capable of producing very *different effects*. A man quite incapable of dead-lifting 4 cwt. may easily hurl a ¼ lb. weight at a velocity which shall make it more than equal mechanically, in some respects, to that dead weight.

5. **Conductors and Di-electrics.**—The resistance of different forms of matter to the flow of electricity varies enormously. Through some the flow seems to be free, as if through tubes or holes; in others it seems held, as liquid in a jelly, wherein globules of liquid are contained within an elastic skin. The first class are called *conductors*, the second class *insulators* or *di-electrics*. In the first, a charge can *flow* through; in the second it has to *burst* through, which only takes place when the potential of the "charge," or difference of pressure, is great compared with the thickness of the di-electric.

But while the distinction is obvious in extreme cases, there is no sharp division. Taking 1 mercury unit of resistance, the resistance in silver and copper is only about $\frac{1}{60}$ part, most of the metals ranging between. That of carbon is about 50 times as much; yet carbon is classed amongst conductors. Acidulated water has a resistance of about 125,000 times as much, but is also often used as a conductor for medical currents. The proportionate resistance of di-electrics like glass, mica, ebonite, is reckoned rather by trillions. Hence their power to "insulate," or preserve, electric charge or pressure. So far as we know, the only absolute, perfect insulator is (or would be if we could get it) an absolute vacuum.

6. **Nature of a Current.**—We may now come to practical points; and as one chief means of obtaining a really steady current is a Voltaic cell, or combination of cells, we will take this in its simplest form, of a plate of copper c, and a plate of zinc z, the latter either quite pure or amalgamated with mercury,[1] face to face in a vessel of some di-electric, such as glass, filled with dilute sulphuric acid. A metallic wire is in metallic connection with each plate. While these wires are separated, if examined by a delicate electroscope the one from c is found slightly charged with positive, and the one from z with negative electricity; there is a difference between them of potential or electric pressure, but nothing further happens.

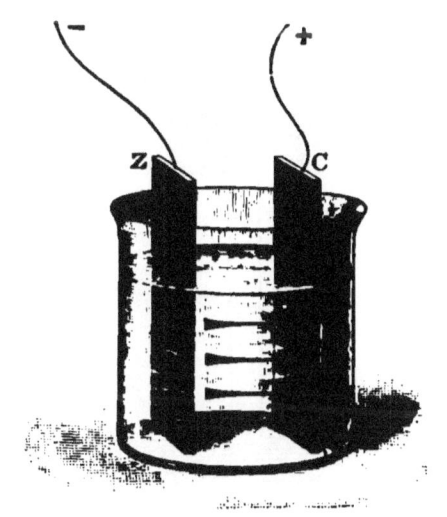

Fig. 1.—Voltaic Cell.

Bringing the wires into contact, however, all is changed. Through the closed conductor now furnished by the wire there is a flow; and if c and z were two *insulated* metal plates with equal charges, the pressure would at once be equalised and there would be an end. As it is, however, there is now also energetic action *in* the cell; the

---

[1] With crude zinc the decomposition does not stop when the wire is disconnected, because two portions of different composition form little couples, which are in metallic circuit. These little couples set up decomposition on their own account, known as local action.

zinc plate is attacked by the acid and water, and torrents of hydrogen gas liberated. By this consumption of zinc Energy is thrown into the current, which *keeps* the copper plate P at a higher, and Z at a lower potential, and a current (which we will call the external circuit) constantly flowing through the wire. Of necessity a current is also flowing *through the liquid*, completing the circuit, and flowing in that part from the zinc to the copper. This is known as the internal circuit. Hence it is, that while the zinc is called the positive *plate* of the

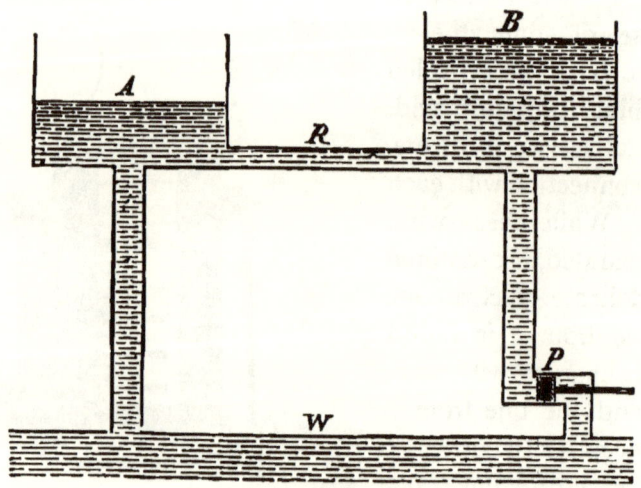

FIG. 2.—A Current of Water.

battery, the copper wire or terminal is called the positive *pole*.[1]

The conception of an incompressible fluid will give us a wonderfully close analogy to all this. Let us have two vessels, A and B (Fig. 2), of which B contains water at a higher level; the two communicating by a channel R, and both communicating with a larger body of water, W, also closed into the system.

[1] The terms are however purely arbitrary. If they really express something of the nature of excess and defect, there are some phenomena which appear to show that what is termed *negative* may be really the *positive*. (See Chapters VII. VIII.)

With nothing further, the liquid would flow down from B to A, till both were the same level, and there would be an end. But suppose a pump introduced at P, which pumps up more water into B as fast as water flows away, then the current is maintained in R. If we further conceive that the mechanical force produced by the current in R is just sufficient to pump up enough water into B, we shall have a complete circuit, as in the cell. It is not sufficient in this case; but in the cell the necessary energy is supplied by the combustion of the zinc.

7. **Combining Cells.**—The analogy of a water current also helps us to understand an important matter regarding the combination of battery cells, for practical purposes, as with an Induction Coil. Suppose we want the water-current R to do some work, and have to double the amount of current which can flow from the one cell B. It is evident that we must employ a second cell, which we will call B_2, of equal content and difference of level; but we may do this in two ways. If we keep the *resistance* at the same amount, by doubling the area of the channel R, we may utilise the second cell at the *same level* as B, and double the *quantity* will flow through R; and so if we want to double our electrical work, but through a *low resistance*, we connect the two zinc plates, and also the two copper plates, which keeps down the resistance in the cells. But if we want to drive twice the water-current through the channel R, still keeping this of the *same* dimensions, the small channel will oppose this by more resistance. We can only overcome that by enclosing the reservoir B, and placing the second one B_2 at the same difference of level *above* B. Then we shall get double the current by doubling the pressure. So also with the voltaic current; if we have to overcome great *resistance* in our work, we must connect the wire from the copper in one cell to the zinc plate in the second; this will double the resistance in the cells, but will also *double the pressure*, or voltage, which is what we want.

The first plan is called joining cells "in parallel"; the second plan joining them "in series."

The practical rule for battery arrangement is got at in this way. Taking six cells, a very common number; if we join all "in parallel" we get only the E.M.F. of one cell, but only *one-sixth* of the internal or cell-resistance of one cell. All "in series" gives us six times the internal resistance of one cell, but also six times the E.M.F. Or we can join pairs "in parallel," or threes "in parallel," putting these combinations into "series." Now the work we want done is the "*external* resistance"; and it can be shown mathematically that we get the largest available current for work, when we group the cells so that the internal resistance of all the cells is as nearly equal as we can get to the external resistance of the work. As Dr. Walmsley expresses it, "when the external resistance is very high the cells should be joined in series; when it is very low they should be joined in parallel; and when it is intermediate a calculation should be made in accordance with the rule."

8. Effects of the Current within the Circuit.—The existence of a flow or "current" in the outer circuit of the Voltaic cell, when the wires are connected, is shown by *effects*. When the wires are not connected there is no effect, except the slight electric charge at each of the two poles. But when the wire is joined, there are several effects, two of which are within the circuit itself.

(*a*) The wire is perceptibly *heated*, if tested by proper instruments. And if the wire be small and of high resistance, like platinum, and the plates in the cell be large, the heat may be so great as to make it red-hot, or even to melt it. The cell itself is also heated. This effect is due to the resistance-friction of the current, and depends upon the resistance.

(*b*) In the cell itself, the zinc is acted upon, and the liquid, which consists of several chemical elements in combination, is *decomposed*. In the cell itself, however, this effect is complicated by other chemical combinations, whose energy it is which is transformed into that of the current; and it is simpler to study this effect in the outer circuit, by bringing into a separate cell of liquid the two wires from the battery. Various compound liquids require different amounts of E.M.F. to decompose them; but if this be sufficient, decomposition takes place in all through which a current will pass, and which are called *electrolytes*. Some complex liquids, such as many hydro-carbons, being di-electric, allow no current to pass, and are, therefore, not decomposed in the ordinary way. This is the *chemical effect* of the current.

9. **Induction.**—But there is another very wonderful effect of the current, in regard to which our water analogy fails us, and we have to fall back upon some such hypothesis as was briefly outlined. Very powerful effects are produced altogether *outside the closed circuit*, or the wire, or any conducting medium whatever, and which are known as *inductive* effects. There are several kinds or phenomena of Electrical Induction, but all really depending upon the production of stresses or strains *outside* the conductor or charged body, in the mysterious entity called Electricity, or, what may possibly be the same thing, in the Ether. And through the Ether permeating them, such strains or stresses are also produced in various forms of Matter which may surround the charged body, or the conducting wire.

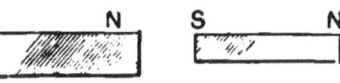

Fig. 3.—Magnetic Induction.

(*a*) The most familiar example of this phenomenon, and the longest known, is *magnetic induction*. Let S N (Fig. 3) be a small bar of soft iron, with no trace whatever of magnetism. Bring near to one end of it, however, the north pole N of a powerful bar-magnet. At once the bar shows all the signs of magnetism, strongly attracting particles and pieces of iron. If it be examined by the usual methods, it is found that the end S, nearest the *north* pole of the magnet, is a *south* pole, while the opposite end of the iron has become a north pole. That the effect is only temporarily "induced" is shown by removing the magnet, when all these phenomena cease. The strain or stress in the space between the two bars is shown by the strong attraction between them; and the definite "lines of force" that surround a magnet are made manifest by the well-known experiment of laying a sheet of paper or glass over a magnet, and shaking thereon fine iron filings. On tapping the sheet the filings arrange themselves into the beautiful "magnetic curves." These curves may be simply described as lines, into which the powerful strains or stresses set up by the magnet strongly twist the two opposite ends of every molecule of the iron, and every large particle which is free to be moved. The intimate connection or identity of magnetism and electricity will appear immediately.

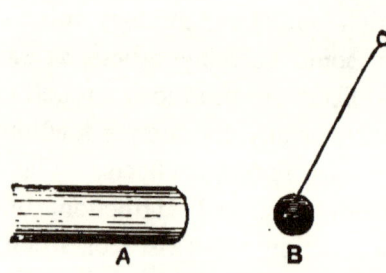

FIG. 4.—Static Induction.

(*b*) Another manifestation is known as *Static electrical induction*. Let A (Fig. 4) be a body "charged" by an electrical machine or otherwise with "positive" electricity, and let us suspend near it, by a silk thread, a pith ball B. On examination by the usual methods the *nearest* side of the ball B will be found to exhibit *negative* electrification, and the farthest side of it *positive* electrification; and the ball B is very strongly *attracted* towards the body A. But it is no vague "attraction" across empty space. All this happens because the two are separated by a *di-electric* mass of dry air; and what really takes place is that a *strong compressive strain or pressure* between A and B is set up in the di-electric. This compressive strain in a di-electric separating two opposite charges of electricity, is experimentally shown to exist by Dr. Kerr's experiments (§ 47).

CONDENSERS

The most important application of this phenomenon is known as a *condenser*, which enables us to absorb and to store large charges of electric energy or stress. A condenser is a thin sheet of glass, mica, paraffined paper, or other di-electric, the opposite faces of which are covered with tin-foil to within some distance of the edges of the sheet. One metallic surface is connected with some source of electricity, generally of comparatively high pressure. If the other face be insulated, the first cannot receive a great quantity or charge; but, if the second surface be connected to earth, it is very different. Supposing the source to charge the first face with positive electricity, this "induces" an equivalent negative charge on the *inner* surface of the tinfoil forming the second face, repelling the positive electricity to its outer surface, whence it passes to earth. This enables more positive charge to be stored in the first tinfoil, and so on through a measurable though minute period till a charge is accumulated depending on what is called the "capacity" of the condenser. This capacity depends upon its surface, the nature of the di-electric, and the thickness of the latter. The thinner the di-electric the greater is the capacity; but this is limited again by the fact that if too thin it may be *perforated* by a spark or disruptive discharge. The di-electric itself is in a state of great strain, which can be made visible by polarised light (*see* § 47).

A familiar form of Condenser is the *Leyden Jar*, consisting of a wide-mouthed jar lined with tinfoil both inside and outside, to within a few inches of the rim. A brass rod with a knob at the top communicates with the inside coating, while the outer is connected to earth or with a source of electricity of opposite kind to the other; thus, the two coatings may be connected to the two secondary terminals of an Induction Coil. For very large condensers, several jars or several di-electric sheets are employed. Their use with coils will appear hereafter.

(*c*) But we are here most directly concerned with what is

known as *Current induction*. A current in a wire also acts, apparently, across space, and "induces" another current in a neighbouring wire; the effect being however really due, as before, to stresses set up in the ether.

We can see this simply and impressively in the instantaneous high-tension current of the "discharge" from a Leyden jar, which was before explained (§ 3) to be of the same essential nature as a voltaic current. On the face of each of two circular plates A and B (Fig. 5) of glass or ebonite, mounted on insulating pillars, let a large flat spiral of covered wire be

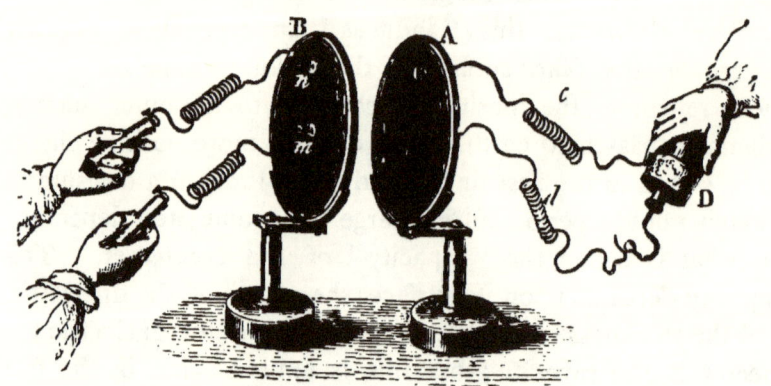

FIG. 5.—Induced Discharge Current.

cemented, using an insulator like shellac; one end of each spiral having a wire from the centre, and the other end from the circumference. Connect one spiral c, with the outer coat of a Leyden jar D by the wire c—the jar will then be discharged when its knob approaches the other wire d, and the discharge-current will flow round the spiral C. Let the wires mn from the other plate B terminate in metal handles, held in the two hands as usual in shock experiments. When the jar D is discharged through the first or "primary" coil on A, a violent shock will be felt from the *induced* current in the "secondary" coil on B.

10. **The Induced Current.**—We want to know more

about this induced current, and a convenient arrangement for affording information is that in Fig. 6. Here E is a voltaic cell, or a battery, from which a current can be sent through a wire A B, with a key interposed at K, by which the current can be sent, or "made" as it is called, and "broken," at pleasure. Parallel to the straight portion A B we arrange the straight portion C D of another wire, whose complete "circuit" goes through the galvanometer G, which will register the *direction* of the current in C D.

What happens is very interesting, and not exactly what any one unacquainted with currents would expect; but we cannot

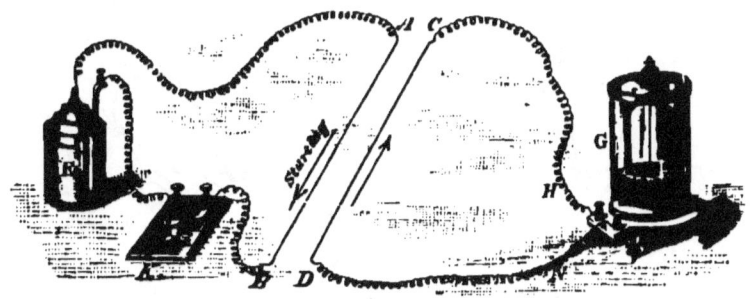

FIG. 6.—Voltaic Current Induction.

discuss the theory of it here. When the current is "made," if it starts in the direction from A to B as marked, a current flows in the *opposite direction* in the other wire, from D to C. Then while the current flows in A B there are no observable phenomena in C D; but when the current A B is *broken*, another induced current flows in the other wire, only this time it is in the *same direction* as A B or from C to D. It should be also noted that if the current A B be *increased*, or the two wires brought *closer together* while current A B is flowing, these changes also produce "inverse" currents; and that if the current be *decreased* in A B, or the two wires be moved *farther apart*, there is a corresponding induced "direct" current.

It is difficult to realise through what an enormous space

these ether-stresses and lines of force really permeate. It may be some help to remember that such induction-currents in straight wires have to be grappled with and guarded against in telephone lines many yards apart; and that it has been found possible to transmit telegraphic messages to the island of Lundy, *some miles* from the mainland, by arranging a long wire parallel to another wire upon the other shore. Messages can be (and are) sent to and from entirely detached and moving railway trains in the same way. The all-important consequences of an electric current are, in fact, the comparatively hidden phenomena in the ether and di-electric matter

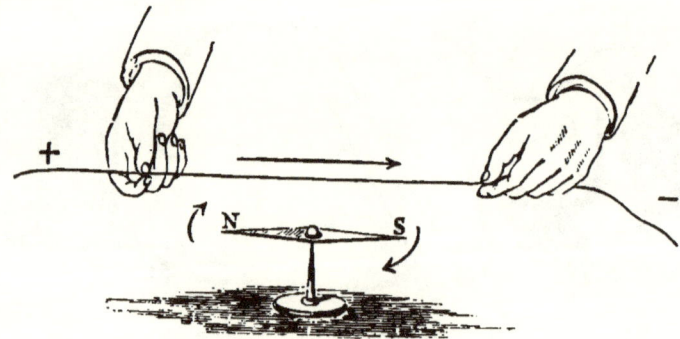

FIG. 7.—Œrsted's Experiment.

surrounding it. The chief office of the wire appears to be to direct and localise the opposite kinds, or polarity, of those phenomena. This, too, is not easy to realise; but we may remember that when a telegraph cable breaks down, it is *not the wire*, but *the di-electric insulation*, which gives way.

11. **Electricity and Magnetism.**—But a Voltaic current has further remarkable inductive effects. The first step towards any real knowledge of them we owe to Professor Œrsted, of Copenhagen. In 1819 he discovered that if we take an ordinary compass-needle N s (Fig. 7), and let it settle in its natural position, and then bring near and parallel to it a wire, through which is passed a voltaic current in the direction indicated, the

compass-needle will be *deflected* in the direction shown in the figure. If the current be sent in the same direction *under* the needle, the deflection is reversed; or a reversed current also reverses the deflection. Thus a wire bent round once, and returning under the needle, doubles the effect. More turns further increase the effect, and in this way we obtain our present sensitive galvanometers.

Other experiments, which need not be detailed, prove that Voltaic currents are always accompanied by phenomena of this "induced magnetism," and that the two are inseparably related, in such a way that the current is always *at right angles* with the axis or line joining the induced magnetic poles. So real and essential is this connection, that if we wind an insulated wire into a helix round a rod, and bring the ends up so that the whole can turn as on a pivot in two mercury cups A B (Fig. 8), to which are led wires C Z, from the copper and zinc plates of a Voltaic cell, the helix (called a solenoid) be-

FIG. 8.—Current in a Solenoid.

haves *in every respect* like a magnet. It turns north and south; if the north pole of a magnet be presented to it, the north pole of the solenoid is repelled and the south pole is attracted; magnetism is induced in a bar of soft iron exactly as by the magnet in Fig. 3; iron filings are attracted; and magnetic curves are produced.

If a soft iron rod be introduced as a core actually within the solenoid, it is powerfully magnetised so long as the current passes. Thus we have an *electro-magnetic induction*. And we also discover, what is very important for our present purposes,

C

that with such an iron core inserted, the magnetic and other effects are *very much intensified*.

12. **Magneto-Electric Induction.**—We should naturally expect, conversely, that a magnet would be capable of inducing an electric current. That it is so was discovered by Faraday. But, just as in induction by a current, it is only "make" or "break," or increase or decrease, or approach or withdrawal of the primary current, which produces an observable induced current; so the magnet must be either magnetised or unmagnetised, strengthened or weakened, approached or withdrawn, to produce observable current effects.

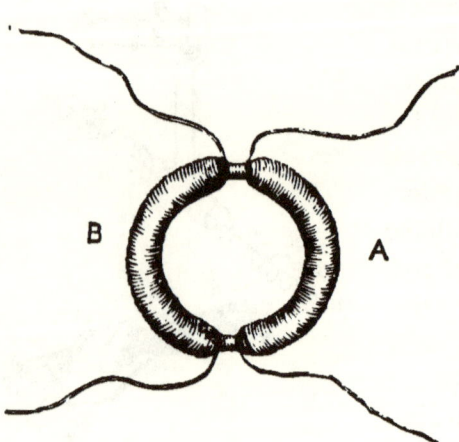

Fig. 9.—Faraday's Coil.

Faraday discovered the phenomenon of magneto-electric induction with the coil shown in Fig. 9, still preserved in the Royal Institution. Round nearly one half of a soft iron ring was wound a helix of covered copper wire A, with projecting terminals, and round the other half a similar coil B. When a current was passed through A from a battery, the ring became magnetised; and, when the current was broken, it was demagnetised. The other coil B was connected with a galvanometer; and it was found that the magnetisation and demagnetisation in the other half of the ring produced current, but in contrary directions. It was an easy step from this to the discovery that, if a coil of many turns be wound upon a hollow bobbin (Fig. 10), and a steel permanent magnet A B, or an electro-magnet, be approached or inserted in, and withdrawn from the centre of the bobbin, the opposite motions produce contrary currents in the wire ff'. Of course it is just the same if the magnet be fixed, and the coil be moved, into the same relative positions. It is in this way that the currents of medical magnetic machines, and the immense dynamos that are now at work on every hand, are produced.

Faraday also experimented with the two coils each wound all round the ring, but the turns of one between those of the other. An iron ring thus wound is a true "induction coil," and is the general type of what are known as "closed circuit transformers" in electrical engineering; because the ring is a closed magnetic circuit with no free poles. He also used such a coil as shown in Fig. 11. Here the battery coil or primary coil B B is wound round a straight soft iron core C; and the secondary coil A A, wound in another helix around B B. This coil is a general type of what are called "open circuit transformers" in electrical engineering, because the iron core has open poles.

FIG. 10.—Magnet and Coil.

13. **Primary and Induced Current.**—This last arrangement contains the main elements of an "induction coil," and it is found that definite features, in the two coils concerned, determine in a definite way the proportions of the current. The pressure or E.M.F. of the current in either is found very nearly proportional to the number of turns; while, on the other hand, of course, doubling or trebling the length of wire (of the same thickness) doubles or trebles the resistance. Supposing then that one coil contains a single layer of thickish

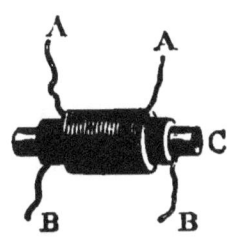

FIG. 11.—A Faraday Coil.

insulated wire, and the other three layers of wire of one-third the section. If we send the battery current through the thick wire, the primary current will be of considerable quantity but low E.M.F., while the secondary or "induced" currents will be less in quantity but of much greater E.M.F. If the battery is connected with the other coil, the current will be "transformed" in the reverse way, as is done in domestic lighting when supplied by alternate-current high-pressure mains. The Induction Coils here treated of, however, are specially designed for the purpose of converting a rather large current of low pressure into small currents of very high tension indeed.

14. **Self-Induction.**—There is still one more phenomenon of induction, specially marked when wire is wound into turns, and which has to be specially considered in the construction of Induction Coils. We learnt that a current "made" in a wire produces a contrary or "reverse" current in a near parallel wire, and another "direct" (or same direction) current when "broken." But *the turns of a coil* are parallel wires. These turns act upon each other in exactly the same way, and if the turns are many the effect is very powerful, since the induced current is raised in tension by every turn in the coil. It will be seen on reflection that both the induced currents are *resisting* currents, tending to counteract the effect of the primary current due to the battery (or other generator). Thus they help to prove that work is done by the direct current, in producing stress in the ether and di-electric matter around; which stress stores up energy like a strained spring, to be given back by a species of recoil. The effects of the "make" and of the "break" are however different.

(*a*) The self-induced current in the coil from "make" is *inverse*, or dead against the battery current itself. It thus while it lasts, *diminishes* the amount of that current, and *retards* the current attaining its full strength. In both ways it perceptibly weakens the induced current in another "secondary" coil.

(*b*) The self-induced current in the primary from "break" is in the *same* direction as the battery current. Thus it prolongs that current after it would otherwise have ceased, or, in other words, hinders the suddenness of its fall; and as the inductive effect, both upon a magnetised iron core, and of the primary current upon the secondary coil, largely depends upon the *suddenness* of the "make and break," this current also is against the final result, or induced current in the secondary.

But there is another very interesting result. The original current cannot exceed the E.M.F. of the battery, and is indeed to a very perceptible degree weakened below this by the opposing action mentioned in (*a*). But the tension of this "direct" induced current depends upon the *number of turns in the coil*, which also increase the resistance; and therefore it may be of quite high voltage, so high as to give a smart shock from a single cell. Hence it is often termed more especially the "extra current," whose effects are easily shown by such an arrangement as Fig. 12. The wire from the battery is coiled (but with many more turns than in the mere skeleton diagram) round an iron core, so as to make of the whole the electro-magnet E M ; there is a key K to "make and break" current; and to the wires on each side of the key-interval are connected a pair of shocking handles *c c*, as on medical coils, one being held in each hand. If only one or two moderate-sized cells are used, nothing will be felt, and as a rule no spark will be seen, on "making" connection. But on "breaking" there will be a brilliant spark at K, and a smart shock through the hands.

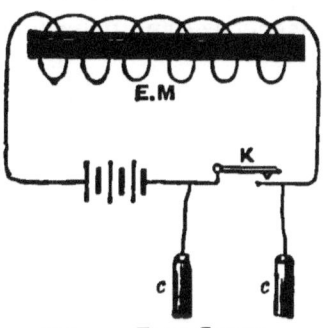

FIG. 12.—Extra Current.

15. **Foucault Currents.**—One particular case, which may

almost be termed one of self-induction, may finally be mentioned. The soft-iron core of the electro-magnet is a good current-conductor. If, therefore, that core be a single massive solid bar, and the electro-magnet be part of a machine (such as a Dynamo or Induction Coil) whose operation depends upon an intermittent current, the "making and breaking" will induce reverse and direct circular currents *round the core*. All such currents, which were specially investigated by Foucault, are (as in the coils) directly *opposed* to the desired or efficient action of the machine. They are guarded against by breaking the continuity of the conduction by splitting the core into separate small wires or thin plates, which are also more quickly magnetised and demagnetised.

CHAPTER II

INDUCTION COILS

THE practical arrangements of an Induction Coil will now be easy to understand. Coils are of several sub-types; but it will be sufficient to take one example of what may be called the low-tension type, and then to explain the construction of the ordinary Ruhmkorff or high-tension coil, as used for the various experiments described in this book.

16. **Medical Coils.**—Fig. 13 shows an arrangement devised at a very early stage by Du Bois-Reymond for his physiological experiments, and which is still in use as a medical coil. In this class of coils, as great tension is not required, and is indeed debarred, it is not necessary to guard against the adverse effects of self-induction, and we find nothing with which the previous chapter has not made us familiar. We have a primary coil or helix B of comparatively thick covered wire, wound round a hollow bobbin, into which a bundle of iron wires can be introduced as a core. One end of this helix is attached to the check fixed on the base-board, and the other end is left free or unsupported. The secondary coil A consists of many hundreds of yards of much finer copper wire, the bobbin upon which it is wound being mounted upon the sledge s. This is known as the "sledge" arrangement, and

its effect is that, with the same current in the primary, the result in the secondary can be varied in two ways; (*a*) by inserting or withdrawing the whole or part of the iron-wire core; and (*b*) by moving the coil A upon its sledge, so that more or less of it encompasses the primary B and is subject to the latter's induction.

Other methods of varying the physiological current are sometimes employed. The primary is in some coils wound in two or more divisions, or two or more layers, with connections by which one or more of these can be cut out of the

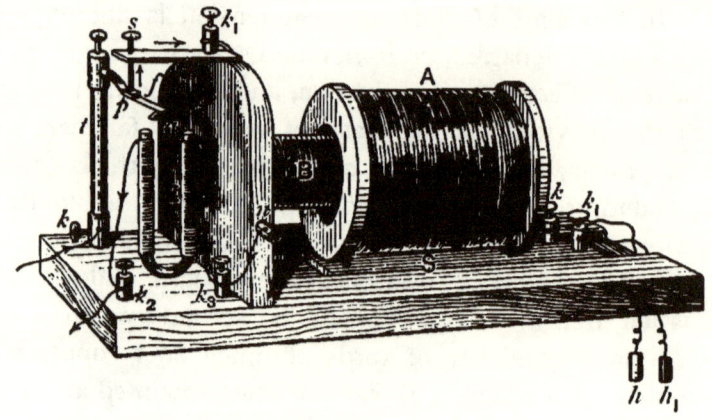

FIG. 13.—Du Bois-Reymond's Coil.

circuit. Another common method is to employ a rheostat of water. Water being but an imperfect conductor, if the secondary current is sent through a column of water some inches in length, the current really available for work is much diminished by the resistance. The water is usually contained in a vertical tube of glass, on the top of which is a cap through which slides a metallic rod, so that the water-resistance can be graduated. It is needless to describe particularly these or other methods of adjustment, which are described fully in the catalogues of medical electricians, and are foreign to the purpose of this work; the more so as induction currents should only be employed physiologically under skilled medical advice. We pass on therefore to other details of more general application.

Medical coils have usually plain screw *terminals*, to which may be attached handles h h_1 for ordinary use, or any other apparatus. The interrupter or contact-breaker is the horseshoe electro-magnet E. The iron armature a is fixed to the end of a spring f which drags it away from the magnet when this is unexcited, thereby keeping the platinum contact-piece p in connection with the screw s. The arm in which s is screwed carries the current to k_1, and thence to one end of the primary B, which the current leaves by n, and thence by k_3 round the magnet to the terminal k_2. The current from the battery thus enters at k and passes by $t f p s k_1$ and round B, thence by n to k_3 and E to k_2, whence it returns to the battery. But this magnetises E, draws down a, and thus "breaks" the current at p; then the armature a springs back and "makes" the current as before. This form of contact-breaker was introduced by Wagner. Its principle is adopted in the majority of coils, but in a much improved form devised by Apps, and shown later in Fig. 16.

17. **High-Tension Experimental Coils.**—For physical and experimental work there are required secondary discharges of much higher tension than would be safe for medical purposes. The potential-difference between the terminals of a coil giving a 1-inch spark is about 50,000 volts, and of a 10-inch spark coil probably about 300,000 volts. The most powerful coil ever yet made, constructed by Mr. Apps for the late Mr. G. Spottiswoode, was capable of giving a 42-inch spark, requiring probably not less than a million volts. Such a spark is a veritable lightning-flash in miniature, though the tremendously long "sparks" of a real lightning-flash must require thousands of millions of volts. In all such coils as we are now dealing with, the prejudicial effects of self-induction in the primary are very apparent, and no great results were

possible, until means were found of modifying or suppressing them.

18. The Condenser.—This great improvement in coil-construction was made by Fizeau, who introduced into a branch or shunt circuit of the primary coil a *condenser* of large capacity (§ 9). This was constructed as in Fig. 14, a number of sheets of thin mica or paraffined paper being interleaved with tin-foil; only it is to be understood that the mica or other di-electric projects all round, say a couple of inches, beyond the metal. Then, as shown in the figure, by projecting strips of tin-foil all the even-numbered metal sheets

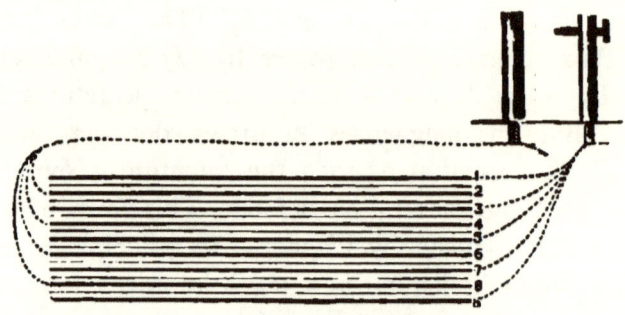

Fig. 14.—Condenser.

are connected to one, and the odd-numbered sheets to the other of two terminals, which are connected with the primary on opposite sides of the spark-gap in the contact-breaker. The effect of this arrangement is that, while contact is "made," the current flows direct through the circuit. When contact is "broken," were there no condenser, the "extra-current" would flow on as a brilliant spark; the current so passing (as we have seen) being in every way prejudicial to the secondary induced current. But instead of that the "extra current" now surges into the condenser, imparting to it a considerable charge, while the spark at the contact-breaker is much less. The secondary spark, or discharge, is on the contrary greatly lengthened.

The reason of this result was discussed by Fizeau, Poggendorff, Faraday, and others, with a tolerable approximation to the truth; but electrical laws and phenomena are more fully understood now than they were then, and there is no doubt at all now about the mode of action, which is interesting as an illustration of modern views about electricity. It will, however, be best understood in connection with a diagrammatic

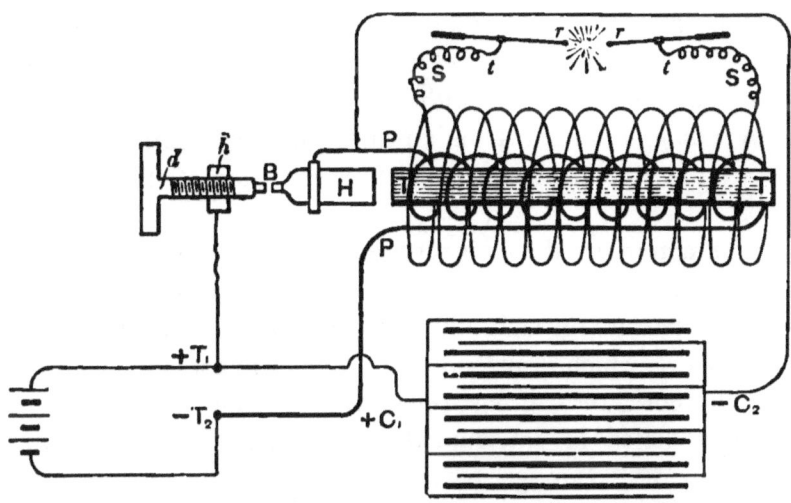

Fig. 15.—Diagram of a Modern Coil.

representation of the entire arrangements of an experimental Induction Coil as made at the present day (Fig. 15).

In this diagram T T represents the core of soft iron wires, often varnished, to prevent more perfectly any Foucault currents (§ 15). The core is enclosed in an insulating tube, round which is coiled the primary coil P P, distinguished as a thicker line in the diagram. This may consist of one, two, or three, or more layers, each turn and layer being carefully insulated. This also is enclosed in an insulating tube, around which are coiled in many layers the turns of the secondary coil s s, shown by the thin line helix, with terminals *t t* connected to discharging rods *r r*. At and near H is shown the contact-breaker. H itself is a cylinder of soft iron, to which one end of the primary wire P P is connected, and which at the end B is armed with a platinum contact-piece; this is borne at the top of a stiff spring (as shown in Fig. 16), so that, when T T is magnetised, H and T are drawn into contact, and, when released, the platinum is drawn back into contact with another platinum contact-piece B on the end of the screw d, which is adjustable in a nut h, connected with one terminal T_1 of the primary battery. The other battery-terminal T_2 is connected direct with the other end of the primary coil P P. Also the terminal T_1 is connected with one coating c_1 of the condenser shown diagrammatically underneath, the other coating c_2 being connected with the primary on the other side of the sparking-gap.

The action of the contact-breaker is the same as in that already described. As soon as the battery is connected the current passes through the contacts at B; but this magnetises T T, which drags away H, breaks the contact, and stops the current; the spring re-establishes contact, and so on. Thus a series of interrupted momentary currents, in one direction, flow through P P.

The action of the condenser can now be readily followed. While the current flows or is "made" through B and P P, no electrical energy can be stored in the condenser, but is stored up as strain in the core T T and surrounding primary. This stored energy is so great, as we have seen, that were there no condenser the potential-difference at the two poles of the spark-gap B would discharge in a brilliant spark. But instead of this the difference of potential is now communicated to the two coats of the condenser, and there re-stored as strain in the di-electric (*see* p. 13). Most of the P. D. at B is thus disposed of, and the spark much diminished, by which (as an incidental advantage) the platinum-contacts are much longer preserved (§§ 24, 101). But the principal effect is that the stored energy of the electro-magnet, not having to overcome the resistance of the spark-gap, leaves P P *much more quickly*; and as the induced current in the secondary depends largely upon the suddenness of the break of current, this inductive effect is greatly increased, and the coil gives a much larger and more brilliant spark when the current is "broken."

There is yet another effect. Energy is now stored as strain in the di-electric of the condenser. But its two coats are in conducting communication through the battery and P P. Hence they again *discharge* through this medium; and it is at once obvious that this discharge-current between the coats of the condenser is in the contrary direction to that in which the re-made battery current itself has to flow. If, therefore, the current is "re-made" at B before this reverse condenser-current has disappeared, as is practically the case, the battery-current has first to overcome this obstructing current before it can produce its effect. Thus the inductive effect of the "make" current is *retarded* by the condenser.

These two effects change in a very important manner the *character* of the secondary discharge from a coil with a condenser. Without a condenser, as in medical coils and Tesla transformers, the discharges are alternately and approximately equal. With a condenser the E.M.F. of the "direct" secondary current on "break" is exalted, while that of the inverse secondary current on "make" is diminished. The consequence is that when the secondary discharge has to overcome much resistance, as is usually the case,[1] only the former current is able to pass. Thus the secondary discharge, instead of being an alternate current, becomes practically *an intermittent current of high voltage in one direction only.*

19. **The Contact Breaker.**—The form of contact-breaker, interrupter, or rheotome, now generally employed for coils

---

[1] Gordon and others have shown that with a small resistance, and using a small battery to a large coil, with very rapid break, equal positive and negative discharges can be obtained. Gordon found this effect with a 17-inch spark coil, reducing the sparking distance to 1 mm.

up to 15 inches or 18 inches of spark, is one patented by Mr. Apps in 1867. As its management should be fully understood, it is more particularly shown in Fig. 16. Here H is the iron hammer, with its platinum contact-piece c, carried at the top of a very stiff spring s, which is firmly screwed at the bottom to a fixed base. A hole is drilled in the spring near its lower end, and furnished with a bearing-collar, against which works a flange on the adjusting-screw N, turned by the milled-head T. This screw moves freely (in an insulating collar m) through the standard A, to which one battery terminal is connected. Opposite H is the other platinum-contact c, at the end of a screw working in A by the milled-head B, with a nut to clamp it when adjusted. Thus the spark between c c has two adjustments — one of length by the screw B, and the other of rate; for by screwing back N by turning T to the left, the spring is tensioned up, requiring a much stronger magnetic pull to drag H away from contact, and thus "slowing" the spark. It is by such a range of adjustment that this, the most convenient form of rheotome, but formerly only applicable to coils of a few inches spark, has been applied to those of a large size; as it is possible by screwing both T and B to the right to make a large coil give such rapid, and consequently small currents, as will only produce half an inch or less of discharge. For powerful discharges the platinum surfaces must be smooth and pretty large (§ 24).

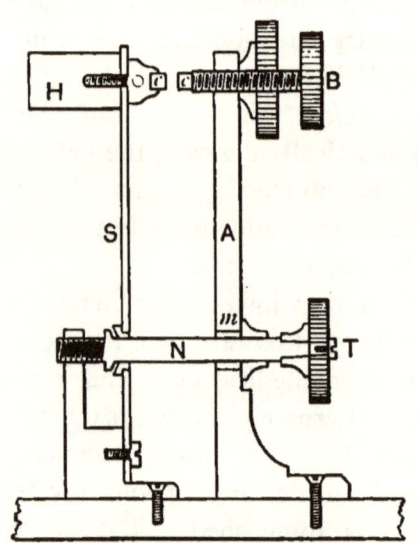

FIG. 16.—Contact Breaker.

There comes a point, however, generally at about 15 inches or 18 inches as regards size of coil, when solid contacts offer too much resistance, and the heavy sparking too rapidly destroys them. In such cases it is usual to employ some form of the "mercury" contact-breaker devised by Foucault. In this form of rheotome a separate magnet is often employed, like E in Fig. 13, actuating a rocking lever, at the other end of which either a single or forked platinum contact enters and leaves a single or double cup of mercury respectively; the fork and double cup are generally preferred. Sometimes the magnet and rheotome are worked by a cell or battery apart from the coil; and sometimes the lever is worked without a magnet at all, by simple mechanical means. Another plan which has been employed is to make the lever in the form of a rectangular bell-crank, the mercury contact-pieces being at the end of the horizontal arm, when the vertical arm can be worked by the magnetised core of the coil in the same way as the hammer of Fig. 16.

The bare surface of mercury would be rapidly oxidised by the sparking. To prevent this, and also check amalgamation of the contacts, the mercury is alloyed with platinum or some other metal, and its cleaned surface covered with some di-electric liquid. Foucault used alcohol, but it is usual now to employ less volatile liquid. If a thin layer of liquid were employed, it would be inflamed and exploded by the sparks: to avoid this, at least an inch in depth of the di-electric liquid is used, and, with powerful coils, much more.

When definite interruptions are required, mechanical rheotomes are used. These may be roughly described as all upon the general principle of a toothed wheel rotated against a metallic spring. More detail is unnecessary, as the operation of any particular form will be understood at a glance.

20. **The Commutator or Switch.**—All coils are furnished with a simple apparatus for connecting or disconnecting the coil with the battery, and also for reversing the

direction of the current in the primary. It may in many cases be more convenient to have the coil itself a little distance in the background, while the switch alone is close at hand. The usual form of commutator was devised by Ruhmkorff, and its mechanism is shown in Fig. 17. A solid cylinder of ivory (v in the top section) has attached to it on opposite sides two crescentic cheeks of brass, v v^1, which each occupy about a quadrant of its circumference, leaving a quadrant of bare ivory between them. Brass spindles a b only enter a certain distance

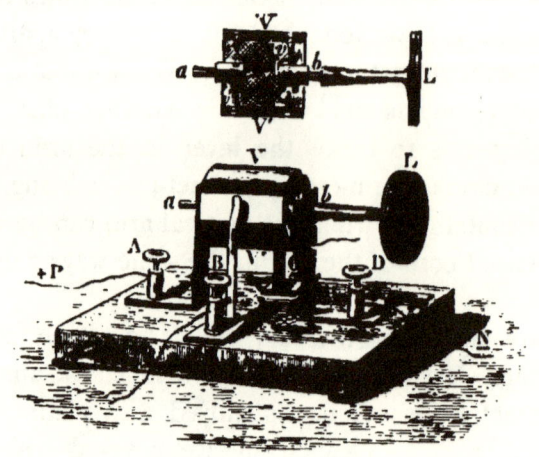

Fig. 17.—Commutator.

at the ends, the ivory also insulating these from each other. The whole cylinder is turned by a milled head L of insulating material; or with large coils it is safer and better to substitute a lever-handle, in order that by its position that of the commutator may be seen at a glance. The positive and negative poles P and N of the battery wires are connected to terminals, A and D, which connect through the metal pillars in which they revolve, with the two end axles a and b; and two opposite points on the sides of the cylinder, by two springs which closely embrace it, are connected with two other terminals B C, which communicate with the primary wire. One of the projecting metal cheeks v is always connected with the axle-end b, and the other v^1 with the end a; in the figure this is done by one of the brass screws v which fix v on the ivory going down, so as to touch b, and another screw v^1 touching a; but another very usual plan is to have a brass disk on each end of the ivory, the cheek belonging to it only extending two-thirds of the distance towards the other end. It does not matter in what exact way the connections and insulations are effected. In any case it will be understood that in the position shown in the lower figure, with the springs B C bearing against the ivory, no current at all reaches the primary. If, on the other hand, L (or a lever-handle) be turned 90° in the direction of the hands of a watch, the current will go through + P A a, v^1 v' to B, and thence to the primary; or, if L be turned to the left, it will go through c to the other end of the primary, reversing the current.

21. **The Discharger.**—All experimental coils are fitted with some form of the well-known Henley discharger, so much used in static electrical discharges. Its original form is shown in Fig. 18. This consists of a small insulated table, generally of ivory, in the centre, for which any kind of holder can be substituted; and on each side of this an insulating pillar of glass or ebonite.

FIG. 18.—Henley Discharger.

To the tops of these pillars are hinged brass sockets, in which slide brass rods furnished with glass or ebonite insulating handles, and which are furnished with screw-terminals for attracting the connecting wires. The rods are screwed on the ends for attaching points or knobs as required, and the sockets should turn in ball and socket joints, though mere hinges will often suffice. Thus by the slides the points of the discharger can be set at any distance within the adjustments; by the ball-sockets in any position; and by the handles slight discharges are effected without shock to the operator. With a large coil, however,—say 8-inch or 10-inch spark—unless the handles are of unwieldy length, the hands would not be sufficiently insulated when the coil is worked at full power to avoid dangerous shocks, since any approach of the hand to the metal portion much nearer than the sparking distance, is liable to draw an irregular spark through the air. Hence the necessity of switching off the current while any arrangements are being made.

Wherever "miscellaneous" experiments are to be frequently made, it is often most convenient to employ a separate and complete apparatus of this kind, and to bring to it wires from the secondary. But the larger coils, meant to be employed solely (or almost so) for discharges in vacua, are usually furnished with the two insulating pillars and discharging rods alone, fixed permanently on the base-board of the apparatus, as part of it. Such an arrangement is shown in the representation of a modern 10-inch coil in Fig. 19. In this case short wires are led from the ends of the secondary to screw-terminals in the discharger. Other screw-terminals are also provided in the latter, from which wires can be led to other detached apparatus, such as high-vacuum tubes. The advantage of this arrangement is that the rods, being set at a proper sparking distance, act as a safety-valve during the experiments in which a high resistance has to be overcome, and which, without such a precaution, and if pushing the discharge to the very utmost, might break down the insulation of the coil. It will be obvious that whatever the resistance of the experiment, or power of the current, the tension cannot exceed the amount determined by the distance at which the discharger is set.

22. **Construction of Coils.**—A coil of 6-inch spark and upwards, as made at the present day, such as is shown from an actual photograph in Fig. 19, depends for its results and pre-

servation upon the somewhat delicate calculation and adjustment of many conditions. The true mutual relations of these, and especially of the all-important one of *insulation* of its parts, have only been discovered by degrees and long experience, and very few makers have attempted anything like exact theoretical calculation, or systematically ascertained, by com-

FIG. 19.—A modern 10 inch Spark Coil.

parative experiments, the best results in points of detail. Naturally, some of the results thus laboriously attained are carefully guarded as valuable trade secrets by the makers who have acquired them. Foremost among these is Mr. Alfred Apps, who has made the largest and most celebrated coils yet constructed, and whose results (compared with the size of the coils) have been so generally superior to those of previous makers that, as far as possible, his proportions are now very usually followed.[1] Taking one of these fine coils as an example, therefore, no attempt is here made to enter into such details, which could only be useful in a minute tabular form not accessible; nor is it indeed desired here to give directions for coil-making by amateurs. Those who take up such experimental work are few; and by them such directions as could be given are already obtainable, while they would occupy all the space available for our more immediate objects. It will only be necessary to explain generally certain principles and methods and details, whose comprehension is important to the successful use and preservation of a good coil, and some of which do not appear to be always understood.

The *core* of a first-class coil consists of a number of separate lengths of the softest charcoal-iron wire, carefully annealed and straightened. The quality is of great importance, in order to quick magnetization and demagnetization. The gauge generally chosen is from No. 20 to No. 22. The straightened wires are generally packed into a brass tube pretty tightly by ramming into the centre a thicker wire or rod which has at one end a soft iron head or armature, on which all the wires are made to abut in contact, and which forms the pole for the hammer of the contact-breaker, a nut at the other end of the rod binding the whole together lengthways. The core is then gradually drawn out from the tube and wound round with a coil of paraffined tape or other insulating material, and finally soaked in melted paraffin-wax until this has expelled all air from the interstices, and formed a further guard against Foucault currents round the core (§ 15).

The *insulating* or di-electric materials of a coil are very important, in all but the smallest sizes. Formerly gutta-percha

---

[1] All his own calculations and results have recently been supplied by Mr. Apps to Messrs. Newton of Fleet Street, who by arrangement now manufacture these coils upon the same *data*, in conjunction with Mr. Apps.

and indiarubber were largely employed, and at a later period resin, shellac varnish, &c. The former gradually "perished," and the latter were found too brittle and apt to crack. The materials now employed in first-class coils are ebonite or vulcanite (the names are synonyms) for the solid parts, and for the construction or winding the hardest kind of paraffin-wax, which has been carefully tested and found satisfactory as regards its di-electric capacity. Silk-covered wire is used for all but small toy instruments.

The *primary coil* has received much consideration. The metal itself is carefully tested for high conductivity, and, to diminish the resistance without occupying more space, a square section was employed at one time. It is still sometimes used, but is tiresome to wind and a little apt to cut the silk covering, and now generally abandoned for round section. For large coils the sizes range from No. 12 to No. 14; below 4-inch spark less is generally used.

Much depends upon the adjustment of the primary to the work required. Every layer of turns increases self-induction and resistance, and with many layers the outer layer is disadvantageously removed from the inducing effect of the magnet, unless the wire used is too small to carry a good current. On the other hand, a greater number of turns increases magnetic effect, which is very important. For special work (or kind of spark) a primary is often specially wound ; and extremely large coils (like that built by Mr. Apps for the late Mr. Spottiswoode) sometimes have it so wound that a number of layers may either act in series as one wire, or as a double or triple strand, of half or one-third the length. For average work, either two or three layers are generally used for coils giving 4 inches to 12 inches spark, and sometimes more for larger ones. The number is partly governed by the proportions of the coil. Mr. Apps some years ago made coils which would give nearly a 12-inch spark, with a secondary bobbin under 9 inches in length. It was justly considered a triumph of insulation to obtain a spark greater than the total length of the mounted coil; but these proportions were afterwards abandoned, and the reasons are not far to seek. Such a short bobbin required four layers of primary, which involves great self-induction; there is also great self-induction in the secondary wire; and, further, such a long discharge above a short coil was very liable to external sparks between the terminals and the outer metal connections. It is found better, on the whole, to wind approximately the same amount of wire in longer bobbins with fewer layers, which cause less self-induction, and keep the wire in a more intense field. It is believed by some electricians that better results might be obtained by lengthening the coil even beyond present general practice, so that the secondary also might be in fewer layers, and within more intense inductive electrical field. But such would involve a longer base-board, and a certain compactness of form is desired by the majority of purchasers.

No special precautions beyond careful and even winding are required in making up the primary coil, as only the extra current can have much tension. The silk-covered wire passes through a bath of melted paraffin-wax as it goes on to the core, and when one layer is wound on, it is insulated from the next by wrapping round it paraffined dry paper. This is done because, while there is hardly any difference of potential between contiguous turns, the length of intervening wire causes considerable difference between two layers, as emphasised presently regarding the secondary wire. Instrumental test is also made as each layer, and finally the whole, is completed. When the primary is all wound on, care is taken by melting and soaking to expel all air, and ensure that the whole length of wire lies embedded in an unbroken medium of the paraffin-wax.

The *condenser* requires careful proportioning to the current with which it is to be used. Strictly speaking, to obtain the very best result, every variation in the length or section of the primary, or in the battery current sent through it, would require some modification of the condenser; as also would the number of layers in the primary. In practice, of course, average currents and general results are studied, and it is better to have rather too much surface than too little. The result of experimental tests is not always what might be expected. Thus, it was found that Mr. Spottiswoode's immense coil, capable of a 42-inch spark, did best upon the whole with no larger a condenser than Mr. Apps habitually used with his 10-inch coils, viz., 126 sheets of tin-foil $18 \times 8\frac{1}{4}$ inches; but extra-separated by two thicknesses of paraffined paper. Unusually large coils are so few, however, that it is very doubtful whether the best proportions, for them, are yet precisely known; because such results are only ascertained by testing to the utmost with various proportions, which is not only costly in itself, but involves too great a risk of breaking the coil down. The di-electric used in coil condensers is a special dry and only slightly sized quality of paper, soaked in the melted paraffin-wax, and carefully examined for absence of any minute perforations.

The *secondary* wire or coil, and its thorough insulation, is the most important consideration of all in coil-construction. If it be considered that the difference of potential between the ends of a secondary giving even 1-inch spark, equals 50,000 volts, it will be seen that there is considerable danger of discharge between such a potential level and the primary, which is of much lower voltage; this necessitates sufficient di-electric insulation between the two coils themselves, in the first place. Such voltage, however, is quite a small matter in coil-making, and a 1-inch coil is often insulated from the primary by merely two or three layers of the paraffined paper already mentioned. This method has been used for 2-inch or even 3-inch sparks, especially for cheap coils of Continental make; but in good coils, from 2 inches of spark upwards, it is customary to employ a tube of ebonite, of calculated thickness, with ebonite flanges at the ends, between which the entire secondary winding is confined.

There are other points to be considered. Taking say a 6-inch coil as a standard size, we may suppose the secondary will require about 6 or 7 miles of wire.[1] We know that about the middle of the wire there can be no difference of potential; but as we recede from this, the difference increases, until at the ends we may get, say 250,000 volts P.D. between them. Merely regarding the primary generally, as at a low potential level, it is obvious that the di-electric insulation between the two *coils* should be thickened towards the ends of the ebonite tube. This can be done by adding on the ebonite tube more and more of paraffined paper, or in any other way.

But, if the wire is to be wound at all closely, which is necessary to keep it within an intense inductive field, the same considerations apply quite as strongly to the insulation of the secondary coil itself. Obviously, between two contiguous turns there is practically no potential difference, and not the slightest (sensible) tendency to discharge. But suppose the secondary be wound like the primary, viz., first one layer all along, then a second layer back again on the top of that, and so on. (Fig. 20.) It will be readily seen that, between the inside layer (representing one end) and the outer layer (which represents the other end) of the wire, there is nearly the whole difference of potentials tending to discharge, separated only by the insulating di-electric interposed between the layers. In particular, the end of wire from the inside layer has to be led out radially towards its terminal, crossing at the end all the layers, and especially the outer one, as at B in Fig. 20.

---

[1] Roughly speaking, if of the best make, these medium sizes require somewhere near a mile of wire for each inch of spark. Small coils can be made with a little less; and as the spark increases the mileage becomes much more, so that Mr. Spottiswoode's 42-inch coil required 286 miles of wire.

No practicable insulation could resist such a difference of potential in a large coil, and such a method of winding is seldom used for anything beyond an inch spark.

The difficulty is got over by a plan, introduced by Messrs. Siemens and Halske, of winding the secondary *in sections*, separated by di-electric disks placed at intervals upon the insulating ebonite tube. Two or four sections are sufficient for small sizes, while a 10-inch coil may be divided into from 50 to 100 sections. Figs. 21 and 22 show how the

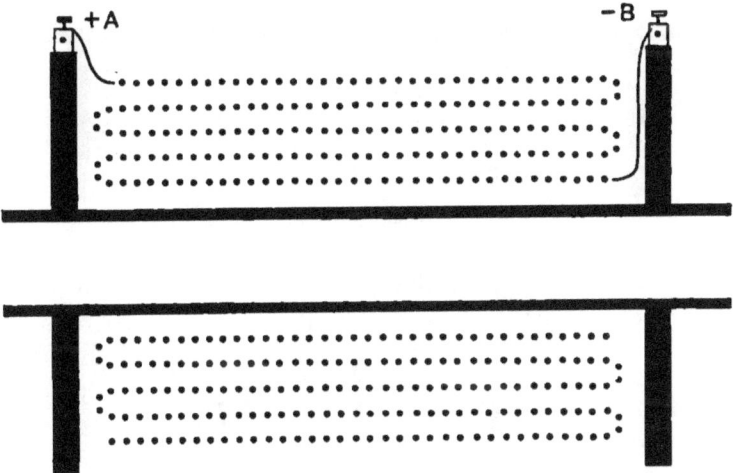

FIG. 20.—Winding in Simple Layers.

system operates, the same letters being used in each, and each *turn* in Fig. 22 standing for an entire double *layer*, as in the diagram above it. It will be perceived that the wire is wound so that the current starts at A in the outermost layer, passing through successive layers to the inside layer of the bobbin at B. There it passes through the di-electric disk B close to the tube, and proceeds from the inner layer successively to the outer one, passing through the disk c at the *outer* edge, and so on to the further end of the bobbin. The sections are always so numbered and arranged in the winding that each end of the coil terminates at the *outer* layer, thus avoiding the dangerous difference of potential shown at B in Fig. 20. To effect this, it will be manifest that the number of sections

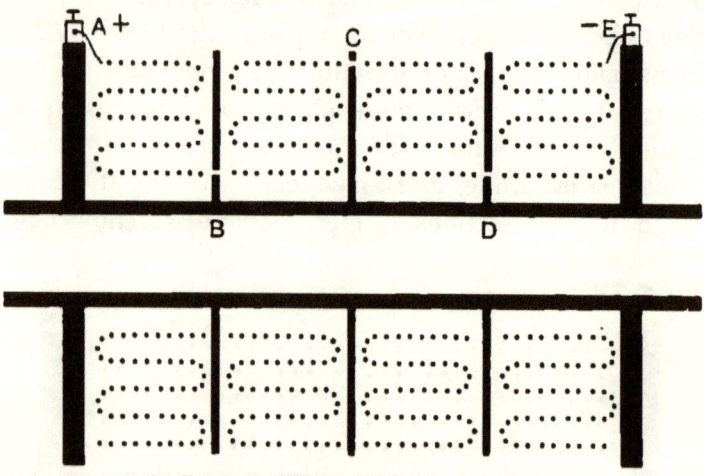

FIG. 21.—Winding in Four Sections.

must always be even. The result of the arrangement is that, if the sections are sufficiently short, there is only a compara-

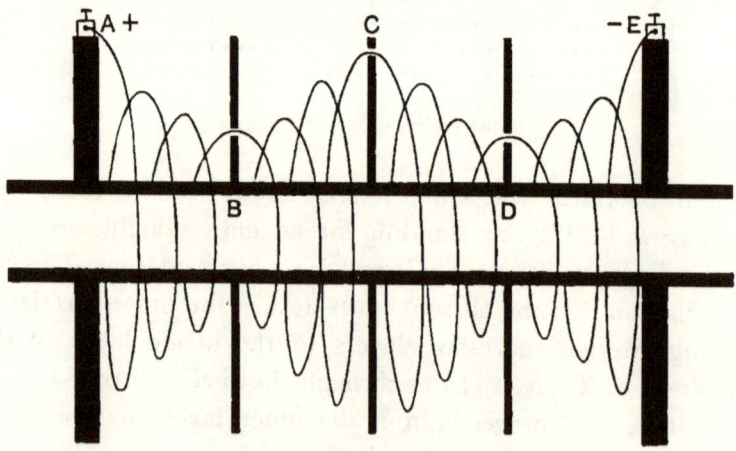

FIG. 22.—Winding in Four Sections.

tively small P. D. between any contiguous sections, which are insulated by a solid di-electric. It is almost impossible to break down the insulation of a coil carefully wound in this way, except by accident.

The width, or rather length, of each section of secondary, depends upon the number of layers. A coil wound so that the completed bobbin is of large diameter, may have each section compressed into a thin disk of coiled wire only $\frac{1}{8}$ or $\frac{1}{10}$ of an inch in thickness; while a long and narrower coil, of much fewer layers, would keep the difference of potentials sufficiently down in a section of an inch or more. Ebonite has been used for the insulating disks; but it is more usual to use several thicknesses of the di-electric paper, joined into one mass like cardboard by melted paraffin-wax instead of paste.

In very large coils indeed, there is another point worthy of consideration. It is evident that the secondary possesses somewhat the character of a condenser, carrying a certain charge, which is more and more concentrated towards the ends. This charge will be more easily carried, if towards the ends the wire be somewhat thickened. This method was employed in Mr. Spottiswoode's celebrated coil; it is rather a disputed point at what size it becomes worth while.

The diameter of the secondary wire depends upon the nature of the spark desired, and the size of the coil. For small coils No. 40 is often used; for larger ones No. 36 is a very usual size; thicker than No. 34 is seldom employed. For the immense coil of the late Mr. Spottiswoode about No. 34 was employed for the central portion, and between No. 28 and No. 30 for the two ends. In large coils it is found best not to wind on the secondary to the same diameter all along, but to reduce the diameter of the outer turns towards the ends. When this is not done, the high potential so near the edges of the end flanges is found very unruly, having a most inconvenient tendency to "brush" and other discharges through the air in all directions.

The secondary wire also is sedulously tested at frequent

intervals during the process of winding, and when finished, great care is of course taken that there is complete insulation in solid paraffin-wax.

Cheap foreign coils are made upon a different plan, which may suffice up to a certain size, but are liable to break down unless understood, and accordingly used with care. The secondary is wound with bare wire only, not touching, but with a space between of about the diameter of the wire, and without any other insulation but the air space. Each layer (such a plan can only be adopted when wound in layers) is insulated from the next by layers of the paraffined paper. Each layer is thus made a section. The principal danger in these coils is of discharge between the outer layer and the end leading out from the inner layer, as at B, Fig 20. The easy way to avoid this would be to make two sections of the coil, as before described, which adds practically nothing to the cost, but brings both ends to the outer layer. I have not happened to see this obvious precaution in the very few coils of this kind which have come under my notice; with it I think a cheap and yet efficient coil might be made on this plan up to about 4 inches of spark.

The *terminals* of the secondary are of course at the discharger, where there are further "screw" terminals from which to lead wires to any other apparatus. The terminals of the primary coil are largely concerned with the condenser, and vary with different makers. An Apps coil, such as here described, has four pairs besides the commutator. The single pair always found on one side of the coil, marked P and N, are for the battery wires. From them, under the base-board, wires go to the commutator, at the edge of the board on the other side. From the commutator wires go direct to the primary, and shunt wires also to one terminal in each of two pairs, marked CC. The other terminal in each of these two pairs goes to one coat of the condenser. By a wire with an insulating handle the two terminals in each pair are connected in usual work, or can be disconnected when this wire is withdrawn. The effect of this arrangement is that *each coat* of the condenser can be either connected or disconnected separately. This is not necessary, as a single pair of terminals would connect or disconnect the whole condenser. But the effect is by no means the same when one coat only is disconnected, as if the whole condenser were thrown out; for the charge still flows into one coat, which acts upon the other by *static induction* in the way before described. The precise effect on the phenomena does not appear very clear: however, all Apps coils are so provided. The other pair of terminals, lettered ALT, or COIL, are connected quite separately and directly to the primary coil only, and are for attaching wires from an alternating "supply" current. When this is used the wires are brought to these terminals *only*, and the current passes direct into the primary, all the other terminals and connections (except the secondary discharger) being out of action. The short wires connecting the pairs of condenser-terminals should of course be withdrawn.

Of course so many terminals and refinements are not at all necessary for ordinary experiments, or with small coils.

CHAPTER III

CHOICE, MANIPULATION, AND CARE OF COILS

The size and character of coil to be chosen will of course depend upon the work to be done, and this should be considered.

23. Choice of a Coil.—For merely pretty experiments, with ordinary vacuum-tubes, a small coil of cheap make, and giving $\frac{1}{2}$-inch of spark, will suffice, and will also demonstrate upon a small scale many of the experiments mentioned in the following chapter. An inch spark coil will perform the same experiments much better, and is quite sufficient when only this class of experiments is intended.

But it is different when either public demonstration of such experiments, or serious work of any kind, is in view. For such purposes a 2-inch spark is the smallest size that can be recommended, and 3 or 4 inches of spark is very much better. Even the 2-inch coil will do much useful spectrum-work, and in the field of Röntgen rays is capable of giving a photograph through the hand in from 80 to 120 seconds. Much of Mr. Huggins' spectrum-work was done with a 3-inch coil; and a 4-inch, with efficient tubes and fluorescent screens, is capable of doing much work with Röntgen rays through the limbs, or even through the body of a young person. But for this latter class of work to be habitually done, as in hospitals, a 6-inch spark is the least size which can be recommended, and medical practitioners usually prefer a 10-inch coil. Very rarely will more than this be advisable, such cases being in fact confined to those who employ themselves with original research into the phenomena of electric discharge, and to whom any advice would be out of place.

Up to about 3 inches, well-made foreign coils are often efficient, at the expense of somewhat greater bulk. Their cheapness has been explained. Above that size, only well-made English instruments can be recommended.

It should also be clearly explained and understood that what is called a 6-inch coil of good English make is one which may be regularly *worked* at the tension of such a spark. The spark that can really be got out of it will considerably exceed that, averaging 8 inches. It would, however, not be safe to work the coil regularly to that point. Foreign and other cheap coils are often estimated at their longest practicable discharge; a discharge which, if constantly used, would probably break the coil down.

24. **Care of a Coil.**—One of the first points in the preservation of a good coil, then, is not to long overstrain it beyond the sparking tension it is intended to bear, which might break down the insulation, and is guarded against by setting the discharger, using a point for the positive pole, and a disk or knob for the negative. This character of the respective terminals must be carefully attended to whenever the primary current is reversed by the commutator. Too large a current in the primary must also be avoided, or the coil may get over-heated and melt the paraffin-wax. The operator should get his hand accustomed to the temperature of about 120° F., and stop work to "rest" or cool the coil, whenever it can be felt that the warmth is approaching that point. It will always get a little warm in use, owing to the resistance in the wire, and a watch should be kept upon this,

For much the same reasons the coil should not be kept, much less used, in too warm a place. This caution should be unnecessary; but it is given because operating-room for the wide-spread Röntgen work is not always easy to find, and I happened to hear of an actual case in which the coil was used in a hot greenhouse, work in which speedily melted the paraffin and broke the coil down.

Coils should be carefully kept out of the damp, and especially out of acid fumes from batteries, and gently and carefully dusted from time to time. When in use dust collects very rapidly, owing to electrical attraction. When damp is unavoidable, and especially sea-damp, Mr. Apps advises to varnish the ebonite with paraffin-wax dissolved in benzol; or in less marked cases to rub over now and then with a little strong soda solution on wool, and afterwards with dry wool.

All the screw terminals should be kept free from dirt or oxide, else proper contact may fail, and the results may suffer. The platinum contacts on the interrupter will also gradually become rough from the residual sparking, and will occasionally require to be smoothed down, and now and then to be renewed. They are really being slowly burnt away all the time the coil is in use; it is only a question of rate and time.

Always cover up the coil with a light case of some kind when not in use, or packed away in its own case.

25. **Coil Manipulation.**—The *current* employed must not exceed the capacity of the coil, and should be adapted to the work being done. A 1-inch coil will be supplied with enough current by a single cell of primary or storage battery; two or three cells may be used for a 2-inch; three or four for 3-inch and 4-inch; while a 6-inch may employ five to eight Grove cells, or four to five storage cells. These will also suffice for a 10-inch coil, or six storages may be used—always with a "cut-out" as advised on p. 55. The electrician will advise

concerning very large coils, or concerning proper reduction of current from public mains.

The *current wires* also claim attention. They must be large enough to convey their current freely; and it is best to use not less than No. 12, using copper of high conductivity. If long wires are necessary they should be even thicker; as, *e.g.*, if a Grove or Bunsen battery be outside the house to avoid the fumes.

A large coil may be employed on what may be called small work, such as exhibiting ordinary small-sized vacuum-tubes. In that case it is better to cut down the current by reducing the battery; but it is always necessary to adjust the tension-spring of the interrupter. The tension-screw T (Fig. 16) should be screwed in till it has almost no backpull upon the spring, when the least pull of the magnet breaks contact, before the full current passes through the primary. The interval between the two platinum contacts will also need lessening. On the other hand, when the fullest and most massive sparks are required, the vibration is to be slowed by screwing back the tension till a very strong magnetic pull is required, and about $\frac{1}{18}$ inch play of the hammer may then be needed; with a large coil sometimes even more is required. This arrangement gathers more energy into each of a smaller number of sparks. With these heavy sparks, special care must be taken to have the discharger terminals properly set, and that the coil is not allowed to become over-heated. When work is over, tension should always be taken off the spring of the interrupter to preserve its stiffness and elasticity.[1]

26. **Personal Precautions.**—An Induction Coil of any size is not to be handled carelessly. Serious and even fatal accidents may occur if the full discharge from a large coil be passed through the body by any careless accident; and as in some branches of work the experiments themselves demand and absorb all the attention, care should be taken from the very

[1] See also § 101.

first to form the *habit and method* of systematic and careful handling. Foremost amongst these habits is that of *never switching on current at the commutator, except and until all is in order for the experiment, and always switching it off as soon as that experiment is completed.*

One element of danger is the uncertainty as to its amount, which mainly depends upon the condition of the heart and nervous system, and may differ incalculably in individuals. Even as regards animals, a rabbit has been killed with a 4-inch coil and single Bunsen cell, and sheep were killed by the great Polytechnic coil. Yet, on the other hand, a rabbit was *not* killed by the latter coil, which gave a spark of 29 inches. This last distance being however much more than the length of the rabbit, much of the discharge must have passed through the air. It is more to the purpose to state some actual experiences. The mere "spark," from a single wire, which may readily happen if the hand approaches too near, is not dangerous unless from a very large coil or to an unusually delicate organisation; but one gentleman records that even such a "spark," from a 10-inch coil, caused violent inflammation in the hand lasting for many days, and involving much swelling, pain, and peeling of the skin. The full discharge from *both* terminals, however, even with only a 2-inch coil, will never be forgotten, and might possibly prove fatal to a weak heart. A very able and well-known experimenter tells us that, receiving (accidentally through a *fall* of some apparatus) the full, though only momentary discharge from an 8-inch coil, he recovered to the knowledge that he had lain on the floor unconscious for several hours.

Bare wires should never be connected with the secondary terminals, and great care be always taken never to approach any metallic terminal with the hands while the current is on. For long and constant experiments it is often well if the coil can be kept in the background, out of the way, while a switch alone can be, if necessary, fitted at hand. But the great thing is a constant, invariable habit of never adjusting anything unless absolutely necessary, until the current has been cut off.

27. **The Current.**—The sources of primary current are either (*a*) Primary batteries; (*b*) Storage batteries; or (*c*) Public supply mains or private dynamo machines. Till lately primary batteries were usually the only available source, and are still very largely used; but in large towns, or wherever a dynamo is available to charge them, storage batteries are far the cheapest and most convenient source of supply. A suitable continuous-current dynamo answers perfectly well. Alternate currents can be used, but do not work advantageously with an ordinary Induction Coil.

28. **Primary Batteries.**—For miscellaneous experiments with small coils, where current is only required for a very short time at once, a large Leclanché or single-fluid bi-chromate cell may be employed, which gives a good current for a few minutes, and will recuperate between the experiments. But much more convenient will be found one of the so-called "dry" batteries, such as the E.S., E.C.C., and others known in the trade, which are all more or less alike in character, and are excited by a paste instead of fluid—hence there is no danger of damage by the spilling of corrosive liquids. Most of them have an E.M.F. of about 1·50 volts per cell, and three large cells will very well work a 2-inch coil for all ordinary miscellaneous experiments. When they are run down, they are re-charged by passing a current through them from a dynamo the other way, so that they have much of the character of storage batteries, though with poles of carbon and zinc. The usual practice, however, is to return them to the makers or agents, who exchange them for charged cells at a specific fee. This system is however only practicable, as a rule, in large towns.

For steady work with larger coils, when primary batteries are used, recourse must be had to one or other of the double-fluid depolarising forms of cell. In most of these the positive element consists of amalgamated zinc, and the re-amalgamation of the zincs from time to time, in order to avoid the zinc being attacked by the acid when the circuit is not closed, is the most troublesome part of battery management.[1] The trouble can be largely reduced, compared with what most people find it, by the simple expedient of using *plenty of mercury* in the process, which costs no more in the long run. The zincs should always be cleaned before amalgamating; with soda, and then rinsed, if new and greasy; with dilute sulphuric acid and a fairly hard brush, if black and corroded. Then pour a good quantity of mercury into a shallow basin, and some diluted sulphuric acid (one part in 6 to 10 of water, or what is used in the cell will do) over the mercury. Introduce the zinc rod or plate into the acid, turning well about, then into the mercury, and "lead" the latter all over the zinc with a stiff brush, till the whole surface becomes silvery white. When all is well coated, the zinc should be well rinsed in water, quite free of the acid, after which superfluous mercury should be brushed off into another basin or saucer with a rather stiff hog-hair brush. If much amalgamation has to be done, it is well to rub a little oil or vaseline over the fingers first, to prevent the acid affecting the skin.

The *Grove* and *Bunsen cells* have in time past been most used for Induction Coils. Each has on one side of the porous division amalgamated zinc, excited by dilute sulphuric acid (from 1-6 to 1-12), while in the other part of the cell the Grove has platinum foil and the Bunsen a plate of carbon, immersed in commercial nitric acid, or often a solution is compounded of this acid and sodic nitrate. The foil is more compact, and the cell has less resistance; but the Bunsen cell is of course far cheaper. Both produce a strong and constant current, but give

---

[1] As *pure* zinc would not require amalgamation (the local action being due to local "couples" between portions of various impurity) it seems strange that it should never have become a commercial product for battery use.

off noxious nitrous fumes, which also corrode the terminals and fittings. Always after use the zincs should be taken out, cleaned with a rather *soft* brush in the same acid as used for the cells, well rinsed, and, if spots appear, re-amalgamated. The sulphuric acid may be used several times, if kept scrupulously from the least mixture of nitric from the other element. The nitric acid may be used as long as it fumes. When it ceases, and looks clear, it is spent and must be thrown away. Six to eight hours is a good average run or amount of work for these cells, by which actual current work is meant.

The *Daniell cell* is rarely used, being bulky and of low E.M.F. One element is amalgamated zinc in dilute sulphuric acid; the other, copper in solution of copper sulphate, with spare crystals of the latter to keep it up. For long Röntgen ray work, however, it is a good battery; running a very long time at a very constant current, with no fumes, at the expense only of some extra space. It will require about double the number of Grove or Bunsen cells.

The constant form of *chromic acid* cell (bi-chromate cells are not so good, tending to deposit crystals) may be described as a Bunsen cell, with a solution of chromic acid instead of nitric acid. It is quite as powerful, and gives off no fumes; and is therefore the cell to be preferred when this type of battery is employed. It is still better to add a portion of potassic chlorate and of sulphuric acid; a good solution being, chromic acid 1 lb., potassic chlorate 2 oz., sulphuric acid 7 oz., water 40 fluid oz. The solution may be used so long as it retains any yellow, orange, or brown colour; but as soon as it has become black or green it is spent. After use the solutions must be at once poured into vessels (if unspent), the cells washed in water, and porous cells *kept* in water, and full of water, till used again. Chromic acid must be most carefully kept from connections.

The only other battery worth special mention is the lately-introduced *Edison-Lalande* cell. This is a single cell, but of great constancy and convenience. The positive element is amalgamated zinc; the negative black copper oxide compressed into plates; the liquid a 25 per cent. solution of caustic potash. The hydrogen liberated, reduces the oxide to metallic copper, and the cell is thus depolarised. A layer of heavy petroleum oil is floated on the solution to prevent "creeping" and evaporation. This cell may be used off and on for months, and gives no fumes and scarcely any trouble. It also has very low resistance, but having also rather low E.M.F., requires a larger number of cells.

29. **Storage Batteries.**—These are much the most convenient where obtainable, but this is only the case where dynamo power is at hand for re-charging, unless double sets are employed, one being sent up to headquarters when run down. Very often even the cost of carriage in addition is cheaper than the running of primaries. One form is well known as the Lithanode cell; another as the E.P.S.; but there are several good and well-known forms, and improvements are constantly being made. Details would, therefore, be out of place, as full directions "up to date" are always furnished with the cells. But one general direction should be made clear. Most cells start with an E.M.F. of about 2·10 volts, which subsequently decreases. Let us say it falls to 1·75 volts per cell (the actual proportion will be given to the user; these figures are liable to alter, as detailed improvements are made). As soon as the E.M.F. falls a certain percentage, the work should cease, because it is injurious to run storage cells entirely down. It is also to be observed that this type of battery is capable of being discharged at very different rates. Thus, if a square foot of surface gives approximately six "ampere hours," this current may be used either as six amperes for an hour, or one ampere for six hours, depending upon the resistance of the work done. Storage batteries should, therefore, always be used with measuring instruments; and the surface (of the single cell, or the number joined "in parallel" as one cell) should be graduated to the amperes required;

coupling the number into "series" which will give the voltage. Also, as any great variation in resistance may suddenly affect greatly the rate of discharge, it is safer with a valuable coil to interpose a fusible cut-out, which will melt and break the circuit when the point is reached that would endanger the insulation.

30. Dynamo Current.—A large coil is very well worked from a continuous-current dynamo or public supply, in the usual way, bringing the adjusted current to the usual terminals. A good average current for an 8-inch coil would be 12 amperes at 10 volts; but the maker of the coil will give the quantity, and arrange the proper adjustments by resistance and otherwise. An *alternate-current* supply can also be used; but the effect differs in important respects. Such a current is brought to the special terminals; or, if there be none such, the condenser is cut out, and the interrupter-contacts screwed up, so that the alternate currents flow *direct* through the primary. These currents being contrary and equal, a condenser would be of no use; and the discharges are equal in all respects, in contrary directions, between the terminals. For lighting vacuum-tubes and some other experiments, the effects of this may be very impressive; but the discharge is *different* from that of an Induction Coil used in the ordinary way. Then, as we have seen (§ 18) by the action of the condenser, one discharge is raised in potential and the other lowered, so that working discharges take place in *one* direction only.

31. Transformers.—The Induction Coil in this latter case, however, becomes in fact a simple alternate-current "Transformer"; the alternating current and cutting-out of the condenser giving it that character. This form of Induction Coil is most used in the form of apparatus so named, "Tesla transformers" being well-known in experimental work. These may be briefly and generally described as induction-coils combining a primary with electro-magnet, and secondary, but without any condensing apparatus (except that such is often introduced into the *secondary* circuit), and the insulation is generally effected by immersing the entire apparatus in heavy paraffin-oil. To go into details of this class of coils is outside the objects of this handbook.

32. Injuries to Coils.—Attention should always be directed to any apparent failure or obvious falling off in a coil. Generally such will need professional repair, but sometimes remedy is possible.

Suppose there is a marked reduction in the spark which can be got between the terminals, while at the interrupter-contacts the spark is much more conspicuous. Such would point in all probability to piercing of the di-electric in the *condenser*. In this case a spare one can be substituted; and then at leisure the perforations in the sheets may often be found and remedied in the first one, which will be ready again to act as reserve.

Break-down of *insulation* ought not to occur if care has been taken. It may be suspected in the *primary* if the interrupter will not work, or if hardly any magnetism can be found in the core. If, on the other hand, the primary coil works perfectly, and yet the secondary spark is much diminished, without increasing the spark at the interrupter, a break-down in the secondary itself may be suspected; especially if the falling-off occurs after straining the discharge to the utmost, or after long work which has heated the coil. Sometimes severe damp will gradually impair insulation. As a rule such accidents necessitate rewinding, or else using the coil only for much smaller sparks; but sometimes careful and persistent heating, to the proper point and that only, will restore insulation. Such heating is, however, a delicate operation, and should be left to the manufacturer.

Foreign dry-wound coils very often need a careful gentle drying through, if the weather be at all damp, to get good results. They should be kept if possible in a specially dry and rather warm place.

CHAPTER IV

MISCELLANEOUS EXPERIMENTS

WHILE a ½-inch spark will suffice for ordinary vacuum-tubes, it is desirable that the experiments in this chapter should be made with a coil giving at least an inch spark. Most of them will be far more brilliant with a larger coil.

33. **Experiments on the Coil.**—Two or three experiments will be very instructive as showing the effect of the "extra current," and the way in which phenomena are modified by the condenser.

(*a*) *The Condenser.*—Adjust the electrodes of the discharger, so as to give the full ordinary spark in the usual way, and then cut the condenser out of the primary circuit. The spark will no longer pass. Let the electrodes be more and more approximated till a spark again passes; that of a 4-inch coil will probably be reduced to about half an inch, and other sparks in proportion.

(*b*) *The extra primary current.*—By connecting to the disconnected condenser terminals, instead of that apparatus, wires with shocking handles, *shocks* will be obtained from the primary coil, as described in § 14. Caution should be used with large coils regarding even this primary shock, which may be unpleasantly smart, using only one cell and little action in the contact-breaker, till it has been ascertained what can be borne.

A more pleasant demonstration, which will take all the primary current, is to connect with the same terminals, instead of shocking handles, an *incandescent lamp*, proportioned to the size of the coil. The extra current when contact is broken will light up the lamp, the intermissions being clearly seen when the rate of breaking current is made rather slow.

These experiments teach us practically that long helices, as sometimes seen, are unsuitable for terminals of a coil, or any other apparatus with an interrupted or alternate current. Suitable enough for battery terminals, and convenient from their elasticity as to length, interrupted currents in a helix lose considerably from self-induction.

Extra-current experiments should be sparingly made, as they injure the platinum contacts more than ordinary work with the condenser in circuit.

34. The Spark Intermittent.—Restoring the condenser to the circuit, we turn attention to the discharges between the secondary. Bringing the electrodes well within the sparking distance, sparks will pass so rapidly as to appear a continuous line; but they may be shown to be momentary and interrupted in several ways.

(*a*). Without any apparatus, by swiftly turning the eyes upward or downward, the sparks will be separated into detached images. Or by rapidly passing the spread fingers between the sparks and the eyes, the shadows of the fingers can be made to give separate retinal impressions.

(*b*). More perfectly, reflect the image of the spark from a piece of looking-glass, and turn this in the hand so as to make the image traverse across it; the sparks will be seen detached. A proper "rotating mirror" will of course demonstrate this still better, and by its aid the images of the sparks can be projected on a screen.

(*c*). Best of all, strengthen the spark by introducing a condenser (such as a Leyden jar) also into the *secondary* circuit, an extra wire besides the sparking electrodes going from each terminal to the opposite coatings of the jar, much as in Fig. 40. The spark will now have much more illuminating power. Prepare a cardboard disk divided into a moderate number of equal sectors, say 10, alternately black and white, which is placed on a rotator to be rotated in the manner of Newton's colour-disk; and placed so as to be seen illuminated by the sparks, the room being otherwise darkened. On rotating, the disk will not appear a uniform grey, but divided into sectors. Further, if the disk has 10 (or n) sectors, it will

FIG. 23.—Disk.

appear absolutely stationary when the time of one rotation is adjusted so as to be 10 (or n) times the spark interval; above or below that rate the sectors will appear to be revolving, either forward or backward. Hence a pretty modification may be made by describing on the disk several circles or zones, and dividing these by a gradually greater number of sectors or spots as the diameter increases (Fig. 23). Then the zone which obeys the numerical law, will appear stationary, while those inside and outside of it will appear slowly revolving, but in opposite directions.

Rotating vacuum-tubes (p. 86) also prove the intermittent character of the discharge.

35. The Discharge in Air.—It is interesting to study the character of the discharge, both with and without a condenser or Leyden jar.

Separating the terminals to the full distance of the coil, and reducing the current by slacking the interrupter-spring T (Fig. 16) till the spark will just not pass, in a dark room there will be seen a faint blue glow round the line joining the terminals. With a powerful coil this "brush discharge" will be fine and conspicuous, and diverging "brushes" more evident at the terminals.

Sparking in the ordinary way, with a point at the positive terminal and a disk at the negative, the spark will be an intense bluish line or succession of lines, with a snapping noise. A sort of yellowish glow surrounding this can generally be distinguished.

As the distance is shortened the spark becomes thicker and more silent, and with a powerful coil appears a sort of yellow flame; with a small coil it is more red. This shorter and redder spark is the proper discharge for all "ignition" experiments.

When a Leyden jar [1] has its two coatings connected with the two terminals, the discharge between the separated points is less frequent, because the jar has to be "charged" before discharge occurs, which may require several of the induced secondary currents. It becomes more massive and makes a loud report; but the sparking distance becomes less, and the less, the larger the jar or other condenser. With a very large jar the spark may be reduced from 10 inches to the $\frac{1}{10}$th of an inch. This kind of spark is often used in spectroscopy (see Fig. 40).

---

[1] Leyden jars or condensers, connected in any way with Induction Coils, should always stand or lie upon an *insulated* table, and never be connected to earth. Neglect of this might lead to very serious shocks. Care should also be taken to *disconnect* entirely the jar or the condenser, as soon as any condenser experiments are concluded.

This effect of a condenser is admirably shown by an experiment of Mr. Hearder's. On an insulated table and surface lay a dry board 2 feet or more square, the upper surface coated with tin foil; and suspend from above by silk or other insulating cords another board, with the lower surface similarly coated, so that it may be raised or lowered over a pully. Each surface being connected up, and one board placed several feet above the other, the points of the discharger will show the ordinary discharge. As the upper board is lowered the two surfaces (probably at 10 to 20 inches apart) begin to come near enough to act as condensing surfaces, separated by air as the di-electric This has obvious effect upon the spark, which at only a few inches distance will be decidedly reduced in distance, and changed in character to that from a condenser.

36. **Effect of Flame.**—The discharger points being set to the full sparking length, introduce the flame of a spirit-lamp (or several lamps in proportion for a long spark). It will be found that the distance can by greatly extended, the discharge from an inch coil being easily extended to 4 inches, and a 4-inch spark to 12 or more inches. This appears partly due to the more rarified atmosphere being an approach to vacuum discharges (Chap. V.), and partly because the vapour is in itself a better conductor.

The latter operation is shown by another experiment. Adjust for the longest sparking distance in air, and introduce one or two flames midway, but on one side of the direct path. Though the path thus becomes longer, the discharge will *diverge* so as to pass through the better conducting flame or heated air.

37. **Deflected Discharges.**—Arranging the discharge at considerably less than the full distance, interpose the edge of a plate of glass, or mica, or ebonite. The discharge will bend, as it were, round this edge, showing that the longer path through the air interposes less resistance than the shorter direct path through the solid di-electric. A plate not too thick may, however, be pierced as in § 46.

38. Current of Air.—Arrange a rather short discharge, and while it is passing blow a stream of air upon it at right angles through a glass tube. The direct spark will still appear, but a curious sheet of flame appears to be driven out from it to one side by the blast of air.

39. Magnetic Deflection.—A still more interesting experiment is to arrange a comparatively short sparking dis-

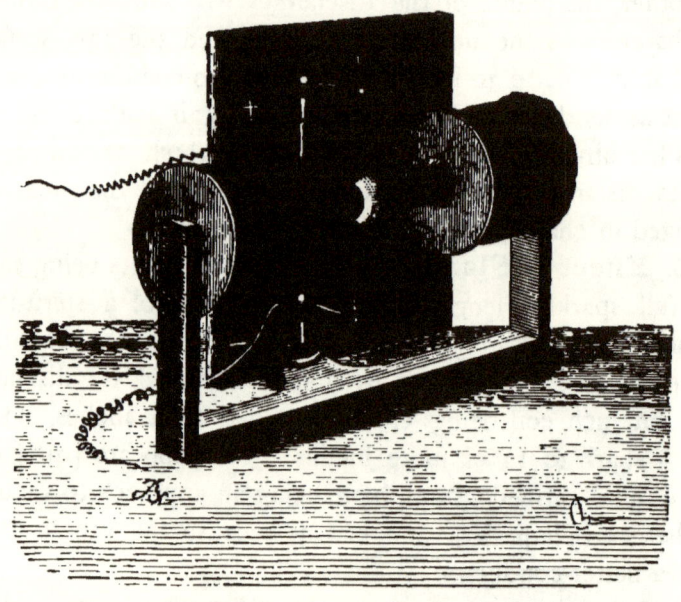

FIG. 24.—Magnetic Deflection.

charge as above, and to adjust across it the poles of a powerful electro-magnet as in Fig. 24; this latter must of course be excited by a branch or shunt wire, or by a separate battery. As soon as the current excites the electro-magnet, the spark is converted into a luminous sheet, deflected into a crescentic shape, from the outer edge of which issue beautiful luminous rays or streaks. If the current is reversed in either the coil or the electro-magnet, the sheet of flame is deflected in the opposite direction.

40. Metallic Sparks and Deflagrations.—Without

trenching upon actual spectrum experiments, if the discharger is furnished with wires of different metals, the sparks will appear of various colours. If there are no terminals to the discharger furnished with binding-screws, thin wires may be coiled round them with ends projecting. A jar in the secondary circuit improves the effect, and will volatilize more refractory metals than the current alone. If very thin wires are used, without a condenser, the wires being of the less perfect conductors, such as platinum, iron, or zinc, it will be seen that the *negative* wire alone ignites, and probably melts, while the positive wire remains unaffected.

That the spark is due to molecules of metal torn off and ignited, is shown by using a point as one terminal, and a polished disk on the negative. At each discharge a small spot will be marked upon the polished surface.

A very beautiful experiment may be made by varnishing the top side of a plate of glass or ebonite, and sifting over it on to the varnish fine filings of copper, iron, and zinc. Lay across each end of the sifted surface a strip or bar of copper, the interval being from three to six times the sparking length of the coil. To these bars the terminals are connected. On discharging, sparks will traverse the entire interval, branching about in all sorts of ways, and of various colours, according to the metals which happen to lie in each particular line of discharge.

All the ordinary sparking effects, obtained with a static machine, are of course available with the coil. One of the prettiest is to "charge" a sparking Leyden jar, which is prepared by gumming on each surface, diamond-fashion, squares of tin-foil, with the corners not in contact, but at a small sparking distance. Each square should have an aperture in the centre, so that the hole in a square of one coating stands opposite to the sparking point on the other coating. The effect of this arrangement is very beautiful, especially if copper or other foils be interspersed, which will vary the colours.

Note.—In *charging* a jar or battery of jars, the arrangement differs from that in which a jar is connected direct to the two terminals as a condenser, merely to strengthen the spark. In charging, there must be a sparking interval between one coat of the jar and one of the terminals. Either one point of the discharger may be directly connected to the jar and the other set at a sparking distance from the other coating; or, what is perhaps better, the discharger is set at a proper interval, and one of the rods connected to its terminal of the coil; from the other discharging-rod the wire is only connected to one of the coats of the jar, while the other coating is connected direct to the other end of the secondary. One discharging-rod is thus only connected to the secondary through the jar or battery, and the whole circuit is interrupted by the spark-gap. In either case, the spark-gap is introduced in order that its resistance may confine the charge to one kind only (see p. 29).

Thin leaf-metal of any kind may be laid and smoothed out on a plate of glass, mica, or ebonite, and the points of the discharger brought in contact with the opposite edges. On passing the current the metal will be brilliantly deflagrated, with the colour due to it, if the coil is powerful enough.

41. **Ignitions and Explosions.**—These are easily effected in endless variety, and the manner needs no detailed description. It suffices to place a little of the gunpowder, or lycopodium, or ether, or phosphorus upon the insulated table of the discharger, and pass the spark through it. The spark will, however, sometimes scatter a powder without igniting it, and gunpowder or lycopodium (the former gently rubbed down very fine in a mortar) should, therefore, be dusted in very small quantity on a tuft of cotton wool.

An extinguished but smoking taper will be re-lighted if the top of the wick be brought between the ends of the discharger, set in line for a short spark.

Small models of any kind of electrical fuse are easily fired by filling a short length of small cartridge-paper tube with the powder or mixture, into the midst of which are led the ends of copper wires thickly coated with gutta-percha, which is removed for an eighth of an inch at the ends. The wires being bound or twisted together, the ends will be separated by a short interval for the spark. Care should be taken with any but a small coil, to lower the current very much for this class of experiments.

Hydrogen or coal-gas, or a mixture of either with air in a brass tube rather loosely closed by a cork, are easily ignited by the spark. But such ignition is most easily and safely shown by bringing a gas-burner, turned on, between the points of the discharger.

42. **The Spark in Liquids.**—Using gutta-percha-covered wires, stripped for an eighth of an inch at the ends, which are brought near together, the spark is easily obtained in water and other liquids. In water, the wires must come within an eighth of an inch for an inch coil. In spirit of turpentine or bi-sulphide of carbon, which are good di-electrics, the spark is very brilliant indeed. In alcohol it is red; in most oils greenish-white. It must be well below the surface in inflammable liquids, and not long maintained, lest the mere heating effect should cause ignition.

43. **Electrolysis.**—Chemical decomposition, though easily effected, is scarcely a legitimate experiment with coils, as the same battery cells alone would usually do the work better. But it is interesting to observe that, while a spark occurs in water when platinum terminals are brought very close together as above, by acidulating the water with sulphuric acid and further separating the poles, and suitably adjusting the discharger, the water is decomposed, and the two gases may be collected as usual. If the conductivity is enough increased by acid and the electrodes near, discharge will take place both ways, and *mixed* gases may be collected over each terminal; while if the resistance is made greater by less acid and a wider separation, the discharge will only take place one way owing to the action of the condenser (§ 18)

F

and oxygen and hydrogen will be separated as by a steady current.

Another experiment is to soak some unsized paper in hot solution of starch, and when dried dip again in rather weak neutral solution of potassium iodide. Lay the wet sheet on a plate of glass, and let the negative terminal (the terminals must be platinum in electrolysis) touch it near the edge. Move the other terminal, which should have a smooth end, slowly over the sheet, and a purple line will be traced, owing to the liberation of iodine and the formation of iodide of starch. Wet litmus paper may be also laid on the glass, and the two terminals brought within half an inch, or an inch, on the surface. A red stain will appear round the positive wire, and a blue stain round the negative. In these experiments only a moderate discharge should be used, and the wires themselves not handled, but each first twisted round the end of a glass rod as a handle, with an inch or two of wire projecting.

44. **Surface Discharges.**—On the table of a Henley discharger, or in front of the coil, lay a piece of looking-glass, silver side downwards, to which is connected one terminal of the coil. Bring the other point of the discharger to the centre of the surface of the glass, whose size should be well within twice the length of the spark. Sparks will diverge from the point of the rod in all directions to the edges of the metal on the back side, and will be duplicated by reflection.

For another pretty experiment, smooth out a strip of tin-foil on the insulated table, over this lay a thin sheet of mica. Bring the points of the discharger to the top surface of the mica, just within the edges of the tin-foil. There will be beautiful branched discharges, and there may be perforations.

A curious variation of this experiment may be made by placing a large drop of water or some aqueous solution on the centre of the mica, or of a glass plate, into which the upper electrode is inserted. The liquid will branch and stream out in a curious way, till a fine arborescent cohesion-figure is produced. This figure will differ markedly in character according to which electrode is dipped in the liquid, and in a less degree according to the liquid, and the di-electric used for the plate.

Another interesting experiment is to dust a plate of glass with lycopodium powder, and lay it on a plate of metal, with the powdered side upwards. The two terminals are then made to touch the powdered surface, at a distance proportioned to the strength of the spark, and a discharge passed. The powder becomes arranged in peculiar figures, which resemble mosses, feathers, and lichens, but are quite different round the two terminals.[1] One terminal may also be connected to the metal plate, and the other brought to the centre of the powdered surface; reducing the spark, or the sparking-gap, so that a discharge does not pass to the edge of the plate. A photographic plate may be similarly treated; but in this case the interrupter should be so handled that only a single spark passes at a time; on development the plate will exhibit characteristic figures, differing according to the terminal used, or the other arrangements, somewhat as in the case of the lycopodium.

45. **Illuminated Crystals.**—Bringing the points of the discharger to a large lump of sugar, or alum, or sulphate of copper, the crystals are beautifully illuminated. With a large coil, drilling opposite holes to some depth in order to introduce the terminals, quite a mass of crystals may be thus treated.

46. **Perforation and Disruption.**—Lay a plate of metal connected to one of the terminals on the insulated table, and upon this a sheet of thick dry paper; and bring the point of the discharger from the other terminal near the top of the paper. The paper will be punctured, and generally appear a little charred round the punctures, so as to make a decided black spot. A sheet held in the stream of sparks, between

[1] See *Proc. Royal Society*, xlvii. 84.

the points of the discharger, will also be perforated, but the holes are not charred, and are almost invisible.

Without any complicated apparatus, the action of Edison's Electric Pen can easily be shown by an Induction Coil. It is only necessary to provide a pencil or stenciller, consisting of a smooth-tipped wire well surrounded with insulating material, such as shellac melted round it in a glass tube. On a sheet of metal, or tin-foil smoothed on a glass plate, lay a sheet of dry paper on which some simple design is traced. Use a single cell, and with a large coil lower the spark, and take care not to let the hand holding the pencil come too near the metal sheet, or a smart shock will probably be felt; but, holding the pencil free, trace with the point over the lines. These will be covered with a row of tiny perforations, the intermittence of the discharge preventing the perforation from being continuous. If the perforations are large enough, some aniline ink brushed over the stencilled design, laid on another sheet of paper, will give a print as in Edison's apparatus.

Plates of glass can also be pierced, but only with a powerful coil; a 10-inch spark might possibly pierce an inch, and almost certainly will pierce a plate half an inch thick. To avoid breaking down the coil, the discharging points must be set at the safe, outside sparking distance, as a safety valve. The glass must be armed on one side with a thick glass tube, ground flat across one end, and well cemented to the plate with shellac. The glass is wiped clean and dry, and the wire from one of the terminals introduced down the tube till it touches the glass, the other wire being brought to the opposite side of the plate. The guard-tube on one side is to prevent the discharge spreading over the surface of the plate.

47. Strains in Di-electrics.—With a large coil may be demonstrated Dr. Kerr's interesting experiments showing optically the strain in a di-electric under a strong electrical charge, but it is diffcult to do this satisfactorily with less than a 6-inch coil. For full details reference may be made to the original papers in *Phil. Magazine*, November and December, 1875, but the essential points are as follows. A block of glass is carefully drilled, and the holes polished, from both sides, till the bottoms of the holes reach within five to eight millimetres. Wires are brought from the terminals to contact with the bottom of each hole, and the interval is arranged in the field of a polariscope, with the Nicols crossed to give a black field, but the polarising planes making an angle of 45° with the line joining the two electrodes. The points of the discharger are first set near together, and the current turned on. The points of the discharger are then by steps (not forgetting to switch the current off while making each alteration) gradually separated. As the electrical strain increases, light re-appears in the polarised field, precisely as if the glass were squeezed in a press, as in fact it is squeezed by the electrical stress. With a distance of about 5 millimetres and a 10-inch coil, Dr. Kerr found a sparking distance of about two inches at the discharger sufficient.

Experiments were also made with liquid di-electrics, such as carbon di-sulphide, with similar results. In this case Dr. Kerr used small flat plates as electrodes, arranged in a manner for which reference must be made to his original papers as above.

If the tension be increased by separating the discharger points (taking care to keep within the safe distance lest the coil be injured) the glass is finally fractured. This usually takes place suddenly; but on one occasion Mr. G. H. Gordon found the final fracture preceded by gradually increasing colour effects, exactly the same in all essentials as when a glass plate is compressed to the point of fracture in a vice.

48. Shocks.—It is not advisable to experiment in this direction with large coils. A ¼-inch spark is as much as is really safe, in case of a weak heart. Such experiments should be confined to small medical coils, or to the primary coil (§ 33).

CHAPTER V

THE DISCHARGE IN PARTIAL VACUA

AT less than atmospheric pressure, the resistance to the passage of the spark or current through gases becomes much less, and gives rise to many interesting and beautiful phenomena. If, on the other hand, the pressure be increased, the di-electric resistance correspondingly increases, approaching more to that of a solid or liquid di-electric. The most convenient mode of studying the simple phenomena is to provide an ovoid glass vessel like Fig. 25, furnished at the bottom with an air-tight stop-cock, and a screw which fits an air-pump and can be placed in a foot. A thick wire with a knob at the tip is in metallic connection with this end, for connection with the coil; and through an air-tight stuffing-box at the top slides another rod furnished with a knob, so that the discharging distance between these knobs can be varied at pleasure.[1]

49. **Spark in Condensed Gas.**—Condensing two or three atmospheres into the vessel, it will be found that the possible spark is much shortened, and assumes a more violent and disruptive character, especially when a jar is introduced into the secondary circuit, as in Fig. 40.

50. **Effect of Rarefaction.**—Set the two sparking knobs of the vessel at twice or three times the normal sparking

---

[1] Such adjustment is not absolutely necessary, but is very convenient and desirable.

distance,[1] and screw the apparatus on an air-pump. Turn on the current, and gradually exhaust the vessel. At a certain pressure the spark will traverse the increased distance, but it still obviously passes as a bright spark, with or without some nebulous brush around it. This experiment is often made with a tube several inches in diameter and two or three feet long, when it becomes a coil adaptation of the well-known "falling-star" experiment with a Leyden jar.

51. **Luminous Discharge.**—As the pressure is further reduced by the pump, the character of the discharge changes, and beautiful streams of purple light (*i.e.*, in ordinary air) proceed from the positive pole or anode; at some given pressure this will nearly fill the vessel. At still higher rarefaction this purple light draws in and brightens into a beautiful reddish-purple spindle, as in Fig. 26, starting from the *positive* pole, while the *negative* knob and rod are clad in a thin blue-violet sheath. When the exhaustion reaches a certain

FIG. 25.—Vessel for Discharges.

point, this violet sheath can be seen to be *separated* from the wire and knob by a very small dark space or distance. The best pressure for these appearances is from 2 to 4 mm. of mercury.

This marked difference between the positive and negative

[1] In an experiment of this kind the discharger should be set as a safety valve at the normal distance, in order to avoid straining the coil.

phenomena in gases is fundamental, common to all gases and all degrees of exhaustion. The higher the vacuum, the more the violet glow (in air) recedes from the negative electrode, termed the kathode. The colours, however, vary with the gas in which the discharge takes place. In most cases the negative glow is blue or violet; but the brush light differs more, and in carbonic acid is pale green.

52. Deflection of Discharge.—If a conductor, or even the finger, be held to the side of the vessel, the brush discharge will be deflected towards it in a curious way. If a strong horseshoe magnet with poles wide apart be employed, curious phenomena will be observed, but which are better studied with "tubes," or as in De la Rive's experiment in § 53.

53. Character and Rotation of the Discharge.—It is clear that the discharge in rarefied air or gas is in a rough way tantamount to a current; but it is obviously of the nature of a *convection* current. Compressed gas is increased in resistance; while we shall see hereafter that even the nearest to a perfect vacuum which we can get, is an all but perfect di-electric. Up to a certain point conduction increases with vacuum; and, beyond that, decreases; which we have no means of accounting for but by freer translation of the particles, up to the point at which there

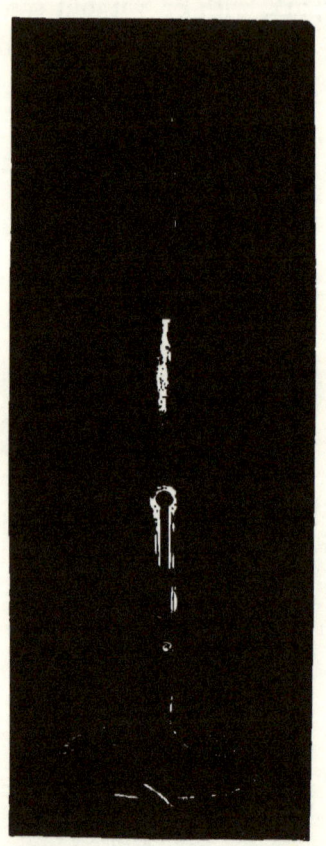

FIG. 26.—Brush and Glow.

remain particles *enough*. Heat may serve as a rough analogy. Heat applied to the upper surface of a liquid penetrates gradually downwards by *conduction*, as it does along a solid body also; applied beneath, it is carried upward and diffused by currents, or *convection*. So also an electrical discharge can travel *through* matter, as along a wire; or it can travel *with* matter, as in this case; molecules of the gas being electrified, and then being carried by repulsion to the opposite pole.

Such streams of electrified particles must obey the same electro-magnetic laws as other conductors, and ought, therefore, to revolve round the pole of a magnet, like the current in a wire.[1] De la Rive demonstrated that this was the case, by the apparatus

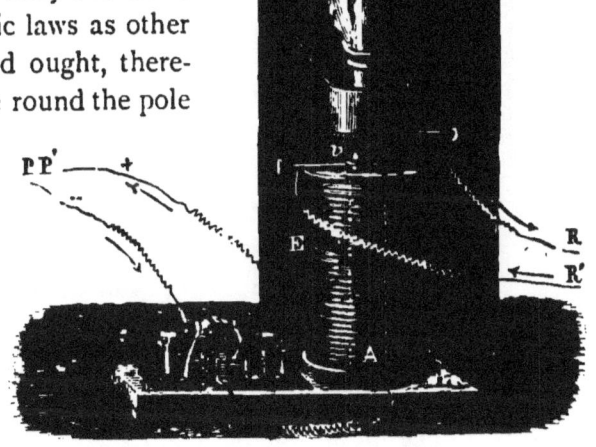

FIG. 27.—De la Rive's Revolving Discharge.

shown in Fig. 27. Into a glass vessel, such as is used in the above experiments and furnished with the usual stopcocks, projects a rod of soft iron A C, whose sides v v are thickly coated with shellac, but the top C exposed. The lower end projects outside, and can be inserted into the coil E, which,

[1] These electro-magnetic rotations are fully described in all works upon electricity. Such details are outside the scope of this little book.

when supplied with current by wires P P¹, imparts strong magnetism to the rod. At the lower end of the vessel, the rod is cemented into an insulating collar, which is enclosed within a metallic collar, and this again is cemented into the bottom of the vessel. Wires R R¹ lead from the iron rod and the metal collar to the secondary coil, so that the discharge in the exhausted vessel passes from the top of the rod C to the metal collar at the bottom. A drop of some hydro-carbon being introduced, and the vessel exhausted, the discharge usually collects as a band at one side; and on magnetising the rod this band revolves round the magnet, in a direction of course depending upon the polarity of the iron and direction of the discharge.

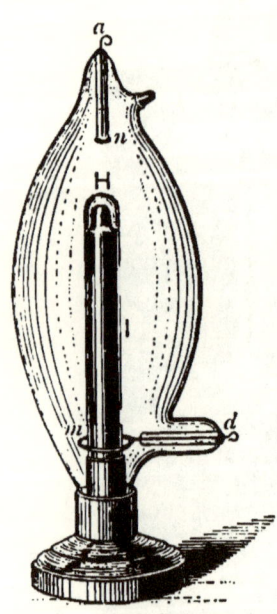

FIG. 28.—Geissler's Apparatus.

This apparatus is needlessly complicated and costly, and Geissler greatly simplified it by sealing into the top of the vessel a wire a with a small disk at the end, and into the bottom a closed tube into which a soft iron rod or a magnet can be inserted. A second wire d is also sealed into the glass, which ends in a ring m encircling the tube. The hydro-carbon being introduced, the vessel is exhausted to the proper point and sealed once for all, in the manner of a Geissler tube; and the discharge takes place between the disk at n and the ring m (Fig. 28).

54. **Gassiot's Cascade.**—Perhaps the most beautiful form of discharge in a low vacuum is that generally known by this name. The effect largely depends upon the scale of the apparatus. In the first form of the experiment, Gassiot coated the inside of a beaker-glass with tin-foil for about two-thirds of the distance from the bottom to the rim. This was placed on an

insulating disk of thick glass, upon the centre of the plate of an air-pump, and over it was placed a wide receiver, through the top of which passed, air-tight, a rod communicating with the tin-foil, but insulated by shellac or a glass tube to nearly its lower end. The positive wire from the secondary communicated with this rod, the other with the plate of the air-pump. When the current was passed, as exhaustion proceeded, a faint glow was first seen about the base of the goblet, which increased in brilliance with the vacuum, until it appeared to pour as a sheet of glowing liquid over the edge of the beaker. The experiment is now generally shown much heightened in beauty by using a goblet of uranium glass, which is not coated with tin-foil; the discharge taking place from a knob at the end of an insulated rod, arranged so that the knob is near the bottom of the goblet. The uranium glass fluoresces strongly by the light of the discharge, and the contrast of its fluorescent yellow-green luminous colour, with that of the discharge, greatly enhances the effect.

A form of this apparatus is often made on a much smaller scale, so as to come within one of the "vacuum-tubes" presently mentioned, permanently sealed, and with ordinary wire electrodes. One electrode terminates in a bulb, which is prolonged into a long tapering glass nozzle, through which the discharge must pass. This reaches to nearly the bottom of a narrow cone-shaped beaker of uranium glass, over the lip of which the discharge has of course to flow, returning through the outer tube to the bulb and other electrode beyond.

55. **Fluorescence.**—The purple light of the brush discharge is richly supplied with ultra-violet rays, and, in comparison with its visual luminosity, has an extraordinary power of exciting fluorescence. This is largely utilised in "vacuum tubes" (§ 57), but can be studied more methodically by exposing cards, painted with various fluorescent solutions, to the rays from the discharge. If the solution be made in one of gelatine, a much thicker layer can be laid on, which will glow

more brilliantly. Solutions in gelatine of uranine, eosin, fluorescein, are very brilliant thus treated. But the solutions themselves are most beautiful, and are easily studied, either by a

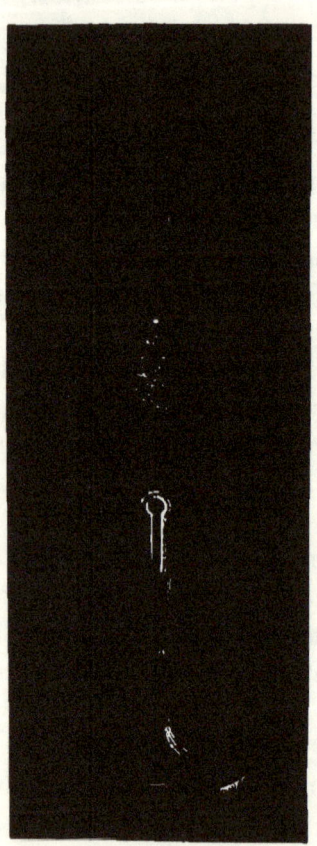

FIG. 29.—Stratification.

"plain" cylindrical vacuum-tube surrounded with a glass collar or annular channel, in which the liquid can be contained, so as to surround the discharge, or in many cases by nearly immersing such a plain tube (keeping the kathode out of the liquid and beyond sparking distance) in a glass goblet of the solution.

Fine solutions treated in this way are quinine di-sulphate (blue), æsculin (blue), heavy petroleum oils (blue), fluorescein and its above compounds (bright green), chlorophyll in ether or alcohol (blood-red). Many other of the more complicated hydro-carbon compounds fluoresce brilliantly by this light, though giving but poor results in ordinary spectrum rays.

56. **Stratification.**—If the exhaustion can be carried far enough, even with common air, another phenomenon will be developed in the discharge; but it is difficult to do so with an ordinary air-pump, and it is easier to introduce by the stop-cock, before exhaustion, a drop of alcohol, or turpentine, or benzol, or almost any volatile hydrocarbon liquid. The reddish-purple brush discharge from the positive pole is then divided into *striæ* (Fig. 29), which increase in number as the exhaustion increases.

57. Vacuum Tubes.—These and many other phenomena are more easily studied in tubes exhausted to various degrees, with the proper gases, vapours, or other substances introduced, and then sealed up once for all with platinum wires also sealed into them at opposite ends as electrodes, to which the wires from the secondary coil can be attached. These were first made in many various forms by Geissler and Gassiot, and are thence commonly known as Geissler or Gassiot tubes.

They can be manipulated in various ways, but one precaution should always be observed—viz., to set the points of the discharger at a sparking distance which is not too great for the tube to bear, otherwise the platinum electrode may be melted, or the tube itself overheated. The primary current itself must also not be too strong. By attaching wire hooks to the terminals of the discharger, the tube may be hung horizontally by its two electrodes, which terminate in small loops, or the tube may be laid (also horizontally) in a cradle or crutch. Or tubes are often mounted, especially double or V-tubes (see Fig. 38), so as to stand upright in a wooden foot. Spiral patterns, such as are shown in the same figure, are usually hung by two small studs upon a piece of blackened board standing upright.

Stratification is best studied in plain tubes resembling Fig. 30, which represents one filled with carbonic acid and exhausted to about 2 or 3 mm. of mercury. This gas, being a compound, shows the stratification more easily than simple gases do as a rule, and the green "brush" or positive discharge is divided into concave disks, the concavity facing the anode. The striæ have been examined under many different conditions; with steady currents from Mr. De la Rive's great battery of 14,000 cells; with intermittent discharges; and with alternate currents from a dynamo and Tesla transformer. When the image of the tube is examined in a rotating mirror, the strata are seen to progress towards the kathode or negative pole. Sometimes an appearance of rotation is observed; but this is generally considered an optical illusion, though it is possible that the magnetism of the earth may really produce rotation occasionally, in the same way as more systematically done by De la Rive (§ 53).

58. **Cause of Stratification.**—The cause and nature of the striæ cannot be said to be thoroughly understood; but on the whole it appears probable that the discharge is not a continuous one, but as it were a series of shorter ones. Schuster further considers the phenomena very analogous to that of electrolysis in liquids, and believes the molecules are broken up into ions, and recombined at regular intervals. These views have since received strong support from some very interesting researches brought before the London Physical Society in 1893 by Mr. Baly.

His experiments were made with tubes measuring about $9 \times \frac{3}{4}$ inches, and exhausted to pressures from $\frac{3}{4}$ mm. to 15 mm., and filled with various mixtures of gases. It has already been remarked that stratification is more easily produced with a compound than with an elementary gas (§ 57). Filling a tube with hydrogen and carbon-dioxide exhausted to $\frac{3}{4}$ mm., when the discharge was first passed, the tube was filled with a white glow, giving (when examined through a pocket spectroscope) the mixed spectrum of both gases. After a few seconds the negative glow became pink, and the white brush light divided into striæ well defined. From these striæ the hydrogen lines gradually disappeared, leaving only the lines of carbon, while

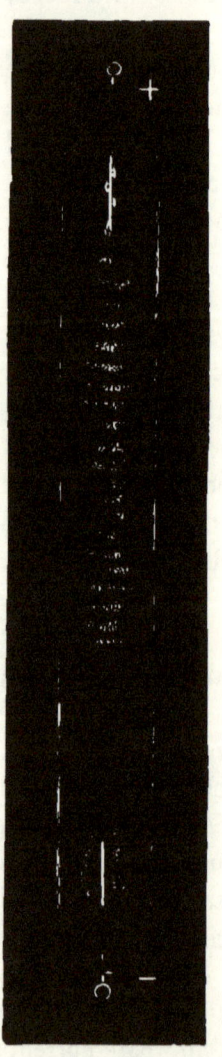

Fig. 30.—Stratification.

the hydrogen lines were found very bright in the negative glow. Generally it was found that, when the mixture included hydrogen, after a while hydrogen lines were strong in the negative glow, but absent from the body of the tube. The inference appeared to be that the hydrogen was, by the action of the discharge, withdrawn from the bulk of the tube, and collected round the negative pole.

Finally, by the arrangement shown in Fig. 31, Mr. Baly succeeded in separating the two gases. The Plücker tube on the right hand was connected with another small tube on the left by a capillary channel which could be sealed off. The upper electrode of the long tube was a rod suspended from the connecting wire by a weak spiral spring, so that its own weight (when in this position) brought it down to touch the other electrode, so making the whole in the long tube one negative electrode, the discharge taking place between this and the lower electrode in the other tube, through the capillary neck. In the small portion of the long tube a morsel or two of cotton wool surrounded the rod, so as to check diffusion. After passing the discharge for some time, the small positive tube was sealed off. On spectroscopic examination of the discharge between its pair of electrodes, this tube showed scarcely a trace of hydrogen, while the other tube, its electrodes being separated by reversing the position, showed brilliant hydrogen lines. Other experiments, in which hydrogen was mixed with carbon monoxide, sulphur dioxide, nitrogen, iodine, &c., gave the same result: the hydrogen was always found at the negative pole.

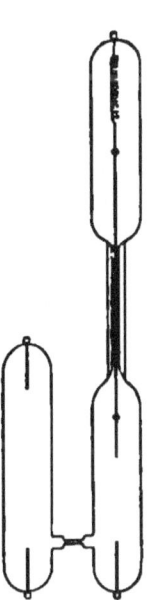

FIG. 31.—Baly's Experiment.

With carbon monoxide and dioxide, the monoxide separated to the negative pole; with carbon dioxide and nitrogen, the carbon to the negative. This mixture separated with special distinctness, and is noticeable for the *heaviest* gas going to the negative pole. Hence the pole does not depend upon the molecular weight. Air, however, is very difficult to determine: nitrogen became very distinct, and apparently oxygen must have been in the other tube, but its spectrum could not be got to resemble the acknowledged type; oxygen is, however, a gas about which there are several mysteries or anomalies not as yet cleared up.

It is clear from these experiments that a process akin to electrolysis is really set up, and that one component can be fractionated out. It was also conclusively proved that good and easy *separation* were connected with good *striæ* in the brush light, and *vice versa*. The first approach of both was always coincident. Again, avoiding any marked negative glow by using for that pole the extreme tip of a wire otherwise sealed in glass, no striæ could be obtained, whereas, on reversing the poles to get a negative glow, they were at once formed. And if pure or elementary vapours are used—as hydrogen, iodine, sulphur, &c.—there are no striæ. The experiments appear to prove conclusively that the discharge has a distinctly electrolytic effect, and that a mixed composition of the residual gas in the tube is essential to the production of striæ. The reason is thus made obvious, why at low exhaustions striæ are so much facilitated by introducing a little vapour of hydro-carbon into the tube.

Plücker devised a form of tube, known by his name, in which the electrodes terminate in end bulbs, joined together by a tube of very much smaller bore—in some cases an almost capillary bore. The discharge is much intensified within the narrow part of the tube, which makes it very convenient for studying stratification (Fig. 32), and also very useful in studying the spectrum of the discharge (see Fig. 41, page 91).

59. Effect of a Magnet.—The effect of a magnet upon the discharges in vacuum tubes is very conspicuous, and manifested in several ways. If the two poles of a magnet embrace the tube between them, stratification is often produced where none was previously visible, and is generally increased if visible before. Again, if the kathode of a tube exhausted so

Fig. 32.—Plücker Tube, showing Striæ.

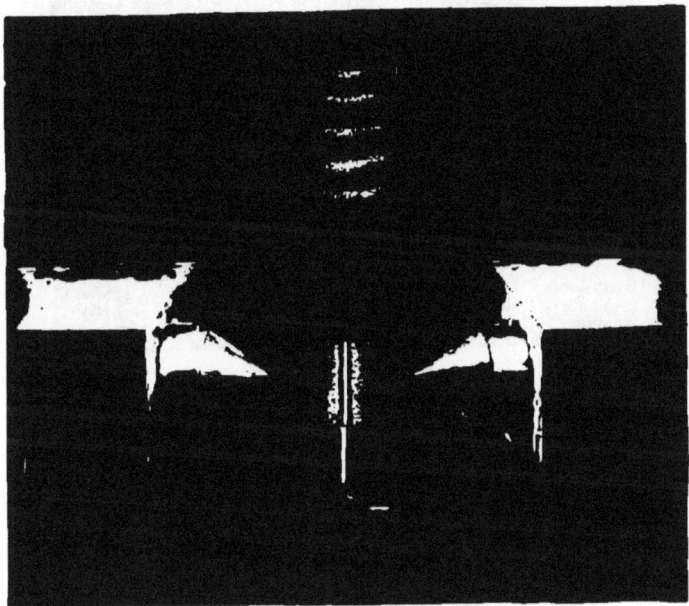

Fig. 33.—Plücker's Plane.

as to show the negative glow well, be brought between the poles of a powerful electro-magnet as in Fig. 33, the glow light arranges itself as if it were a paramagnetic body, in a plane between the poles, known from the discoverer as Plücker's plane. The positive brush-light, on the contrary, behaves nearly as if a diamagnetic body; and if the exhaustion be carried to the right point (usually from 0.5 to 1.0 mm.) can be shown nearly at right angles with the Plücker plane. Plücker demonstrated further very interesting phenomena in the paramagnetic negative plane. A tube of the form shown in Figs. 34, 35, 36 was exhausted to a point that showed

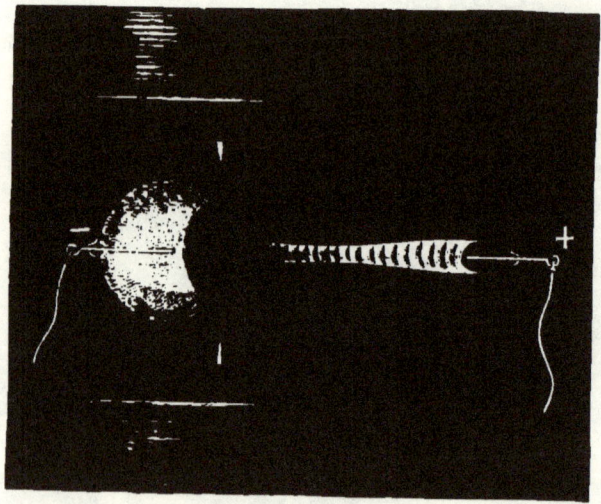

FIG. 34.—Plücker's Curve.

the negative glow well, while the positive brush discharge displayed broad and distinct stratification. Fig. 34 shows the form assumed by the dark boundary of the plane, or negative glow-sheet, when the line joining the poles of the magnet crossed the negative electrode at some distance beyond its point; and Fig. 35 shows the reversed curve when also crossing the electrode, but some distance below the point. When the axis of the tube lies over and across the poles, the tip of the electrode being midway between them, the phenomena are as in Fig. 36. It will be observed that the curve nearly coincides with the *magnetic curve* or "line of force" between the two

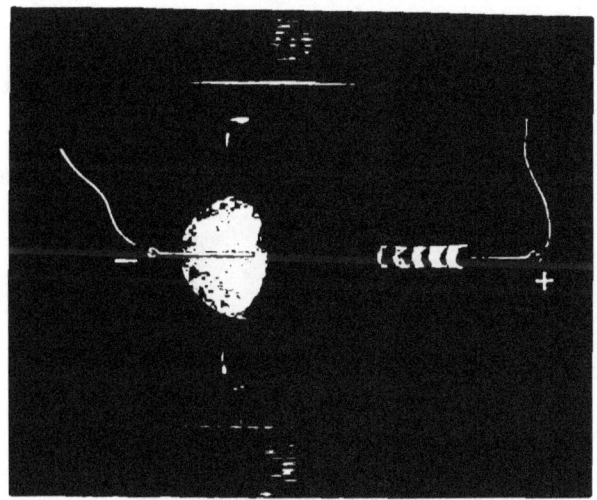

FIG. 35.—Plücker's Curve.

poles of the magnet, and passing through the end of the negative electrode.

Another very interesting magnetic effect was discovered by

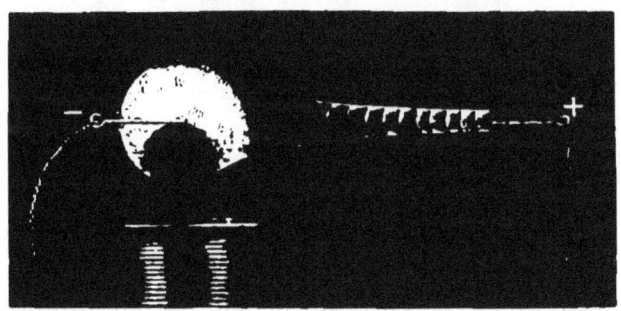

FIG. 36.—Plücker's Curve.

Trèves, and cannot be said to be thoroughly understood, though no doubt connected with the foregoing phenomena. Taking an ordinary Plücker tube, like Fig. 41, he arranged this as in Fig. 37, between the poles of an electro-magnet of great power. Passing the discharge, an exhausted hydrogen tube exhibits in the capillary portion a conspicuous red colour. Switching on a current to the electro-magnet, this colour is changed to white. An oxygen tube similarly treated changed from white to red;

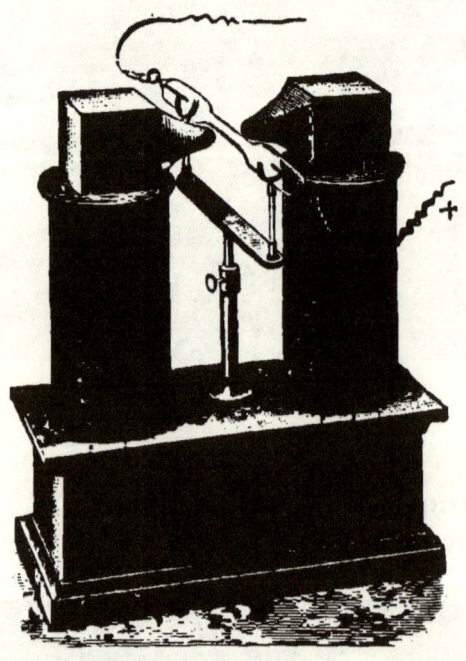

FIG. 37.—Trèves' Experiment.

nitrogen from pale blue to dark blue; carbonic acid from white to blue; chlorine from bluish-white to deep blue.

60. **Display Tubes.**—For purposes of display, vacuum tubes are manufactured in almost endless variety, a few forms only being shown in Fig. 38. Simple patterns from 5 to 6 inches long can be purchased for as little as 2s. each, while very large ones, 2 or 3 feet long, specially made for powerful coils, and of complicated pattern, may cost 2 or 3 guineas. The inner tubes are sometimes made of fluorescent glass and

have the discharge sent through them, the large outer tube being merely a protective case ; or more frequently the discharge passes mainly in the outer tube, while the smaller ones inside are filled with fluorescent solutions of diverse colours. It is needless to describe this class of tube more in detail. In yet another modification, the discharge is passed through a central comparatively narrow combination, which alone is ex-

Fig. 38.—Vacuum Tubes.

hausted, while the large outer tube is filled with fluorescent liquid.

61. Gassiot's Star.—A beautiful effect is produced by attaching several tubes in the position of radii upon a whirling apparatus, connection with the coil being made by bringing wires from their electrodes either to the opposite insulated ends of an axis, or to an insulated metal centre and metal ring, both at one end. The rotation may be produced by hand as in Fig. 39, which shows three tubes so mounted, or by a small electro-magnetic motor. The latter is most usual, but hand-rotation is really the best, because the speed can be varied.

It will easily be understood that if the tubes were illuminated by a steady current, there would be an unbroken disk of variously-coloured light; but as they are only illuminated instantaneously at intervals, the effect is that of a wheel with many spokes, whose number depends upon the speed of rotation. The beauty of the total effect depends far more upon a judicious combination of various colours in the tubes, than upon any great elaboration in their detail.

62. Phonoscope.—A further and yet more beautiful modification may be introduced. If the interrupter or contact-breaker be screwed up, so as to pass a steady current through the primary, and there be introduced into the battery circuit instead, a Reiss transmitter or some other form of telephonic apparatus, which is sung into, the "make and break" will no longer take place with

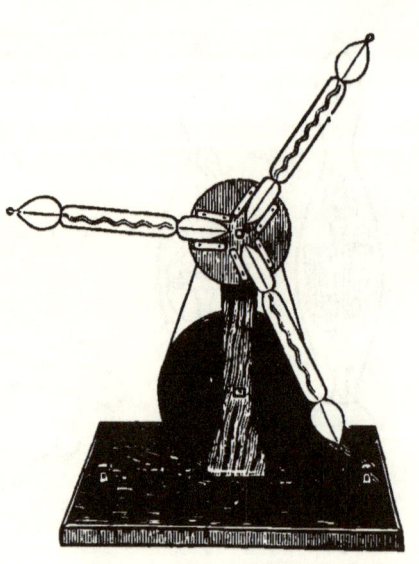

FIG. 39.—Tube Rotator.

regularity, but in very various manner according to the effect of the voice. The consequence will be constant variation in the pattern of the star. These effects may be made of great beauty if the combination of tubes be carefully studied, and are still more varied if one of the terminals on the rotating axis, instead of a continuous metal ring always in conducting contact, be split up into two insulated segments somewhat like the commutator of a dynamo machine, to one of which some tubes are connected and to the other the remainder, while a spring or brush is at one period in contact with one segment, at another with the other, and at other times with both.

[It may be well to state that ordinary vacuum tubes are best exhausted to about 2 to 4 mm. of mercury. Such tubes, of small size, are very well shown by a coil giving a $\frac{1}{2}$-inch spark; and most of the ordinary "effect" tubes are well shown by an inch spark, to about which point a large coil should be reduced. Experimental work may require much more, and such effects as Gassiot's Cascade, it has been already remarked, increase in magnificence with the scale of the apparatus.

Ordinary vacuum tubes will also work perfectly well with the direct alternate current from a public main, if a transformer be at hand which gives about 0·02 ampere at 5000 volts.]

CHAPTER VI

SPECTRUM WORK

THE details of this class of work must of course be sought in the text books upon spectrum analysis; we are only here concerned with a brief description of the principal appliances and methods used in connection with the Induction Coil.

63. **Spectra Composite.**—The spark between the terminals of an Induction Coil giving a tolerably bright light, it is obvious that if we arrange it in front of the slit of a spectroscope we shall obtain its *spectrum*. In the case of all discharges which can fairly be called "sparks," this spectrum is found to be of a very composite character. Part of it will be due to the ignited vapour of the metal (vapourized by the heat of the spark), and part to the ignited gases of the air or other medium in which the spark is taken. Included in the spectra will be the lines of any *impurities* in either of the metal terminals or the gases.

If the spark is very short, the metal lines will predominate in the spectrum; the longer the interval, the weaker will become the metal lines, and the brighter will become the lines of the gaseous medium. The spectrum of air will contain the lines of nitrogen, oxygen, and (from aqueous vapour) hydrogen, and it will be found that this spectrum is brought out better with terminals of some metals than with others.

64. Sparking Arrangement.—Metals of which the spark is to be examined are fashioned into electrodes, usually pieces of wire, and fastened into insulated clamps to which the secondary terminals of the coil are connected, as shown at the right hand of Fig. 40. If a number of spectra are to be compared, a number of clamps are sometimes arranged round a disk which can be revolved, so as to bring a number of lower terminals in succession under one common upper terminal, arranged in front of the slit of the spectroscope; the average distance for sparking metals will range between 3 and 6 mm.

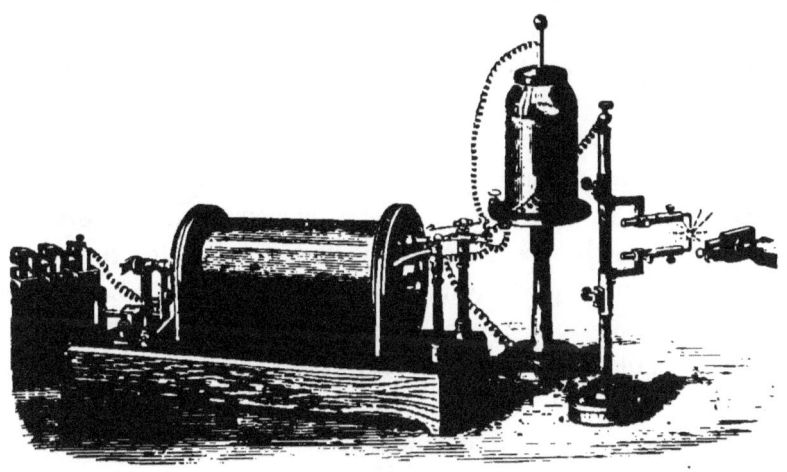

FIG. 40.—Sparking Arrangements.

But it is necessary, in any critical observations, to remember that vapour from one metal will travel through the atmosphere sufficiently to contaminate a neighbouring electrode of another, and mingle the first spectrum with it; hence in many cases the second or other following metals must be kept quite away from the immediate neighbourhood, and only placed in the clamp when about to be sparked.

The spectrum of the common electrode is of course known. Electrically-deposited pure copper is often convenient. Other electrodes which have been used, are platinum, gold, or alloy of tin-cadmium or lead cadmium, which give strong and convenient reference lines, easily compared with and distinguished from the other portions of the spectrum.

65. Temperature and Spectrum.—The temperature of the spark has a great deal to do with the character of the spectrum; and when a higher temperature is required than can be had from the ordinary spark, a Leyden jar or other form of *condenser* is connected by a shunt wire with the terminals, as also shown in Fig. 40. A flat plane of glass, coated on both sides, known as a "Franklin pane," is often used. The condenser must stand upon an insulated table of some kind.

With either arrangement—that is, either with condenser or without—the spectrum of the spark between metallic terminals usually consists of bright *lines* upon a background of less bright *bands*; and with the condenser the bright lines become more numerous and complicated, most of all when a coil of the largest size is employed. At low tension the lines, broadly speaking, are the spectrum of the metal or metals; the bands that of the air or gas in which the spark is taken. Lead gives many beautiful violet lines; zinc a splendid green band or group of lines; gold chiefly in the yellow and violet; silver many green lines.

But there is no fundamental distinction between lines and bands; though it may be said broadly, that bands are as a rule to be regarded as the spectrum of the vapour of *compounds*, lines of pure or simple elements. The conversion into lines at higher temperature, is of course connected with this; the compound being at higher temperatures more and more dissociated and broken up. Yet again, bands are more or less connected with *pressure* of the vapour, so that even hydrogen, for instance, under great pressure, will give a spectrum of more and more banded character, till at very great pressure it may be continuous. But as the pressure diminishes, the banded character of a gaseous spectrum also diminishes, until at last, especially with high tension sparks, it becomes one of bright lines.

66 Plücker Tubes.—The spectra of gases at various pressures is best studied in a form of vacuum-tube devised by Plücker for the purpose, and shown in Fig. 41, two bulbous ends being connected by a much smaller bore, which may be nearly a capillary bore. The discharge is much brightened and intensified in the narrow part, and thus gives a much better spectrum. For demonstration, a number of tubes are prepared with different gases at different pressures; for investigation, the method of course is to connect a tube with the air-pump, and examine the gas or vapour at various pressures or degrees of exhaustion.

Yet another method of investigation is to throw an *image* of the spark actually upon the slit of the spectroscope, by a separate achromatic lens. This enables *different portions* of the spark to be passed across the narrow slit, with the result of giving very various spectra, with differing lengths of lines, according to the differing state and composition of the vapour at different points.

Spectrum-work with a coil therefore consists in examining the spectra of discharges of different solids, or of the same known electrodes in different gases or vapours, at differing electric tensions, different gaseous pressures, and either as a whole, or by separating different minute portions of the small spark.

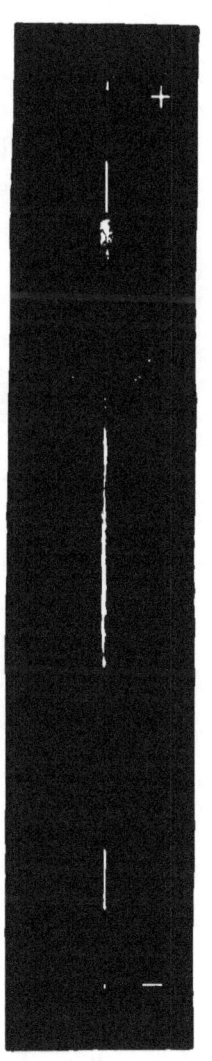

FIG. 41.—Plücker Tube.

Further details cannot be given in a work of this character.

67. Spectra of Salts.—Salts of metals are examined either in the solid form, or in solutions. Powdered salts are generally pressed into a conical cup made in the end of a thick wire, which is drilled with a small hole up the centre or otherwise, through which a small platinum wire rises up the centre of the cup like the wick of a candle. The central wire, whose use was introduced by Mr. Friswell, centralizes and steadies the spark. The powdered salt is pressed as tightly as possible into the cup round the wire.

For solutions, some form of apparatus called a *fulgurator* is used.[1] A very simple and effective apparatus recommended by Mr. Lockyer is shown in section in Fig. 42. Here A is a test-tube through the bottom of which is sealed a platinum wire terminal *f*. Through the cork C at the top end passes a capillary tube B, in which passes centrally the upper platinum wire electrode *cd*. At D is a small bit of almost capillary glass tube, slightly tapered, which when dropped over the lower electrode rises barely above it. The solution or other liquid to be sparked is dropped into the tube, but always keeping it *below* the level of D, as at *ab*. It then rises by capillarity between the wire *f* and tube D, collecting in a little cup, as it were, over the top of the wire. Thus its quantity and level are kept uniform, and it will go on sparking until exhausted.

FIG. 42.—Fulgurator.

68. Phosphorescent Spectra.—There is yet another branch of spectrum analysis, which in the hands of Mr. W. Crookes has lately assumed considerable importance. Its origin was in the intense phosphorescence excited by molecular bombardment, in the high-vacua experiments

[1] Several forms are figured in Schellen's *Spectrum Analysis*.

described in the next chapter. The *spectra* of the bright phosphorescence thus produced in various bodies were soon found to be exceedingly characteristic, especially in the case of certain rare earths; and by pushing this method of examination, in connection with the chemical operation of fractionating solutions, Mr. Crookes was enabled to separate the one supposed oxide of yttrium into oxides of no less than five rare metals. At a comparatively early period in these researches he stated, "it has been my practice to submit all these anomalous bodies to molecular bombardment."

Details of Mr. Crookes' work and methods in this branch of spectroscopy will be found in the *Philosophical Transactions* from 1883 to 1885. Generally, it will be sufficient here to state that the usual method was to submit any particular product, as arrived at by fractionation, to phosphorescence in a highly-exhausted bulb much resembling Fig. 46 in the next chapter, but very often connected with the air-pump so as to obtain the best degree of exhaustion. The electrodes often consisted of tin-foil disks *outside* the bulb.

CHAPTER VII

THE DISCHARGE IN HIGH VACUA

WHEN a vacuum-tube is exhausted to a very high point indeed, ranging, let us say, from one-millionth of an atmosphere to even ten or twenty times such a rarefaction as that, an entirely new set of phenomena make their appearance, different in many respects from those described in Chapter V. One of the most cardinal and obvious differences is that, whereas the phenomena of ordinary vacuum-tubes are concerned chiefly with the positive or "brush" light of the discharge, in very high vacua it is the *negative* phenomena which are so conspicuous. But experimental study further shows that the rarefied gas now behaves in such a very distinct and remarkable way, as to give ground for the belief that besides the ordinary three states familiarly known to us as solid, liquid, and gaseous, matter exists in a fourth state, which may be called "radiant," as different from the gaseous in its physical properties, as that is from the solid or the liquid. Faraday with extraordinary sagacity uttered a prophecy that it would probably be so, in lectures so far back as 1816 and 1819.[1]

Certain points about these phenomena have been investigated by Hittorf, Puluj, Dr. Goldstein, and others, besides Mr.

[1] *Life and Letters of Faraday*, i. 308.

William Crookes; but it is to the latter we really owe the bulk of our knowledge of them, their best explanation, and most of the experiments by which they have been so far elucidated. Even at this date it is not possible to present a clearer popular outline of this part of our subject, than by a summary of the impressive series of experiments and researches brought by Mr. Crookes before the Royal Institution, and again before the British Association, in 1879; and it is best to do so, wherever condensation does not compel otherwise, in his own sequence and method.

69. **Mean Free Path of Gaseous Molecules.**—Gases are believed to be composed of an infinite number of small particles or molecules, constantly moving in every direction with immense velocity, the actual velocity depending upon the temperature of the gas. These molecules are so numerous, that no one can move far in any direction without coming into molecular collison.[1] But if we exhaust the air or gas in a closed vessel, the number of molecules being so much diminished, the distance through which any one of them can move without coming into collision with others is increased, and thus the length of what is called the *mean free path* of the molecules is inversely proportional to the number of molecules present. As this mean free path is by higher exhaustion more and more prolonged, the physical properties of the gas are more and more modified, until at a certain point the phenomena of the *radiometer* become possible. On pushing exhaustion still further, the phenomena occur now to be described.

70. **The Dark Space.**—In describing the phenomena of even ordinary vacuum-tubes, mention has already been made of a small "dark space," which appears to separate the negative "glow" light from the actual kathode or negative pole (*see* Figs. 26, 32). Mr. Crookes very early connected this dark

---

[1] It is probable that molecules of matter can by no force known to us be forced into absolute contact in space.

space with the mean free path of the molecules, and found that it increased or diminished as the vacuum was varied, and the mean free path lengthened or contracted. This result can be very readily shown by a Fleuss or mercurial air-pump in connection with a tube, but is made more conspicuous by such a tube as shown in Fig. 43, where there is a pole in the centre used as a kathode, in the form of a metal disk, and wire poles at each end, which serve as a duplicate or divided anode. There will then appear a double dark space in the centre, on each side of the kathode; narrow if the ex-

FIG. 43.—Dark Space.

haustion is low, but at a tolerably high exhaustion extending for an inch or more on both sides of the kathode. It is terminated by a very bright boundary, supposed to occur at the distance where the negatively electrified molecules, repelled with enormous velocity from the kathode, come in conflict with the molecules moving in their various paths at the boundary of the dark space, and there by collisions produce luminous effects.

71. **Phosphorescence.**—By carrying the exhaustion still further, the mean free path and consequent dark space may be extended so far as to reach the containing tube itself, with

results which are very impressive. The molecules do not now have their direct motion arrested until they reach the surface of the solid glass; and this bombardment by them upon solid bodies at once reveals the extraordinary power radiant matter possesses of exciting *phosphorescence.* Of this effect the long mean free path appears the only tenable explanation. We have no longer to deal with matter in any *continuous* form, as we have even in gases, at natural pressure or only moderate exhaustion; but the molecules are able to exert their own

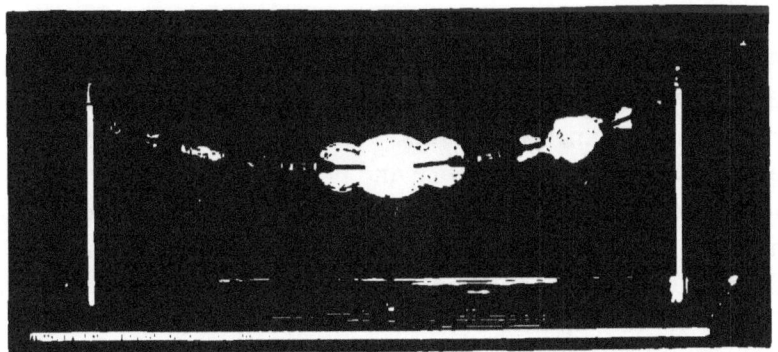

FIG. 44.—Phosphorescence of Tubes.

action as projectiles, and the effect of their own motion and momentum, individually.

It appears to be this direct *impact* of the molecules which produces such powerful phosphorescent effects. Fig. 44 shows three short or bulbous tubes whose electrodes are connected in series, so that the discharge passes through them all. The tube *a* is of uranium glass, and phosphoresces dark green; the tube *b* is of English glass, and phosphoresces of a blue colour; and *c* is of a German glass (a kind of which many various tubes are commonly constructed), which phosphoresces a very bright apple-green.

This *phosphorescence* must be distinguished from the *fluorescence* of many vacuum-tubes described in Chapter V. Fluor escence is ordinarily an effect produced by light-rays, or invisible rays of the ultra-violet *light-waves*, which set up in the material vibrations of another (generally a longer) period. In most cases this effect ceases almost instantly the exciting rays are withdrawn, but not quite instantaneously; and so far as the effect really is prolonged afterwards, the phenomenon is linked with phosphorescence, with which undoubtedly the connection is very close. But in this case no proper light-rays excite the luminous effect: it is the rapid *bombardment* by molecules which excites the luminous vibrations. There is another difference in the greater permanence of the luminous effect: when the current is cut off the phosphorescence can be seen to very gradually fade away, and, in the case of certain substances, it remains for hours. No sharp line can, however, be drawn between fluorescence and phosphorescence. In many substances, such as uranium glass, both phenomena are very marked; and both may be excited by true light-rays, as when Balmain's paint is excited by the rays of the sun. The common feature to both is that the luminous effects seen are not due to light *reflected*, but to light *emitted*, owing to the molecules of the fluorescent or phosphorescent body being themselves excited into luminous vibration.

A vast number of substances, which cannot be excited (or at least sensibly so) to phosphorescence or fluorescence by any light-rays, are powerfully affected by the molecular discharge in high vacua. Mr. Crookes found the diamond the most ready and sensitive substance of all, and one of his most impressive experiments was the bright phosphorescence of a large stone mounted in a highly exhausted tube, at the centre of curvature of a concave or cup-shaped negative pole, the reason for which latter arrangement will appear directly. This particular diamond phosphoresced a brilliant green (Fig. 45).

A large diamond will be out of the reach of most experimenters; but tubes are easily obtained with pieces of coral, or certain minerals of a not expensive character, similarly mounted, which will phosphoresce very brilliantly. The ruby is perhaps nearly equal to the diamond; and as many small stones are of little value, while all appear to phosphoresce equally well, tubes containing a number of small crystals loose, as in Fig. 46, can be obtained for a moderate sum (averaging 15s. to 20s.). The phosphorescence is a bright, rich red, and this is of about the same tone and colour, however deep or

FIG. 45.—Phosphorescence of Diamond.

pale be the "natural" hue of the stones, some of which are nearly colourless.

There is more to be said of singular interest about this ruby phosphorescence. A ruby is crystallised alumina with very little colouring matter, and pure prepared alumina is a white powder. Nearly forty years ago Becquerel published a paper describing the behaviour of such prepared alumina in his "phosphoroscope"—the white powder glowed with this very same rich red. If we prepare a Crookes' tube with precipitated alumina, under this different treatment by molecular bombardment it also shows the same rich red, and if examined by a pocket spectroscope is found to give the same spectrum —especially a strong red line a little below the line B—which is also seen on examining the red light reflected from a good ruby. When the discharge ceases, after a while the alumina resumes its former appearance. But Mr. Crookes records that,

FIG. 46.—Phosphorescent Rubies.

after the same tube had been repeatedly used in demonstration, *i.e.*, when the alumina had been long and often exposed to this molecular bombardment in the tube, the white powder began to acquire a *pink tinge*, which gradually deepened, until the characteristic ruby spectrum could be discerned in the unexcited alumina. It thus appears as if the repeated impacts had gradually shaken the alumina into a different molecular arrangement, perhaps even caused some approach towards crystallisation.

72. **Exhaustion and Phosphorescence.**—There is, as we should expect, one particular degree of exhaustion the most favourable for the production of phosphorescence; and this was found to be roughly about one-millionth part of an

atmosphere.[1] Beyond that the effect begins to weaken perceptibly, until a point is reached when the discharge refuses to pass. This is demonstrated by such a tube as is shown in Fig. 47, where the two electrodes are at a and b, and at c is sealed on a small supplementary tube communicating by a small aperture, and containing some solid caustic potash. The tube being exhausted to a high point, the potash is heated to drive off moisture, and this exhaustion and heating is repeated many times, after which the tube is sealed up. The tube being connected with the secondary of the coil, when the current is turned on, at first nothing is seen, the exhaustion having been carried so far that no discharge can pass. Warming the potash

FIG. 47.—Potash High-Vacuum Tube.

slightly, however, expels a trace of aqueous vapour, when at once conduction commences, and green phosphorescence flashes out along the tube. Driving out by heat still more vapour from the potash, this phosphorescence diminishes in favour of the more cloudy luminosity seen in lower vacua; in this, by and bye, stratification appears; and finally the discharge passes as the ordinary purple "brush." The lamp being withdrawn, the process is reversed as the vapour is gradually re-absorbed, until the green phosphorescent light again appears, and ultimately this also should disappear.[2]

---

[1] Corresponding to 1·1300th of a millimetre of mercury in the barometer; that is, 1 mm. pressure is 1316 millionths of an atmosphere. An average common "vacuum-tube" is about 4000 times this pressure.

[2] The tubes usually sold to exhibit this experiment very often fail to act satisfactorily after the first time. Very long and tedious continuance in alternate heatings and exhaustions before the tube is sealed are necessary for anything like permanence; such exhaustion greatly increases the cost, and is hardly possible for a commercial article.

73. Radiant Molecules Move in Straight Lines.—The absolutely straight lines, in which we are considering the electrified molecules to travel when repelled from the negative pole, may be demonstrated in several striking experiments, some of which also show that this path is always a normal to the surface of the kathode. Fig. 48 is a V-shaped

Fig. 48.—Inability to Turn a Corner.

tube furnished with an electrode at each end of the V. When the pole at *a* is made negative, or kathode, the whole of that arm glows with the green phosphorescent light, but none of this can turn the corner to get to the other arm of the tube. If the current is then reversed, so that the other pole becomes the kathode, the green phosphorescence is at once transferred to the other arm, while the positive arm becomes dark in turn.[1]

These features, it will be seen at once, stand in marked contrast with all the phenomena of ordinary vacuum tubes. In the latter the luminous effects evidently start from the positive pole; however convoluted the tube may be, the discharge readily follows all its contortions, and the luminous effect is seen to reside in the *gas* contained in the tube, from which any fluorescence in the glass itself, due to the powerful ultra-violet rays, can be readily distinguished. But the phosphorescence caused by the molecular bombardment in a high vacuum is *entirely independent of the position of the positive pole*. Let us test this by comparing the behaviour of two tubes exactly similar in every detail (Fig. 49), except that the one to the left is exhausted to about 3 mm. of mercury, whilst the other approximates to the millionth of an atmosphere. Each tube is furnished with a concave kathode a or a', whilst at top, bottom, and side opposite the kathode, it is supplied with three wire anodes, either of which can be connected with the positive pole P of the secondary coil. In the case of the ordinary exhaustion, on the left hand, as we shift the connection from one anode to another, the violet line of discharge changes from one to another of those shown in the figure, always proceeding from the actual *anode* that happens to be employed. But with the other tube it is totally different. The molecular streams or "rays" proceed normally from the concave surface of the *kathode*, crossing at the centre of curvature as at a "focus," and, thence diverging, fall upon the opposite side of the tube and cause a green phosphorescent patch of light. If now we change the anode from b to c, or to d, it

[1] Further experiments in connection with these "kathode rays" (*see* Chapter VIII.) have shown that, if a tube be bent at several right angles, the phosphorescent patch at each angle is capable of emitting or reflecting "rays," which excite some phosphorescence at the next angle, and so on. These effects, however, stand apart from the more vigorous phosphorescence excited by the direct bombardment of the gaseous molecules.

makes no difference, the phosphorescent patch still remains where it was, and nothing at the anode is observable at all. Where the positive pole is, appears of no consequence whatever, or at most of very little indeed.

It will further be observed that the molecular streams

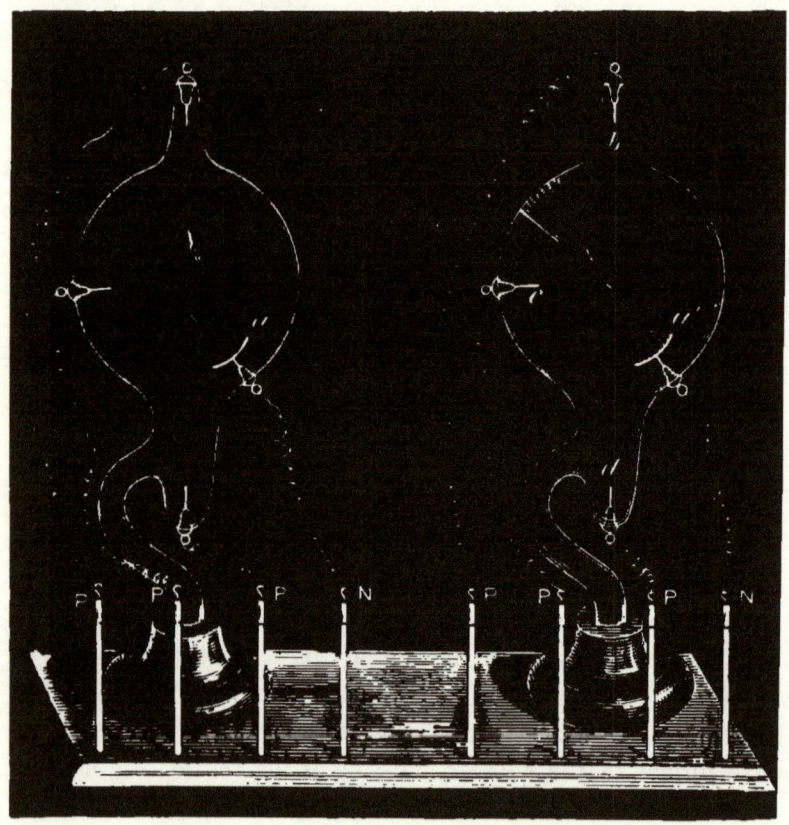

Fig. 49. Low and High Vacuum Tubes.

proceed, as just now remarked, strictly at *right angles*, or normally, from the surface of the kathode. Thus they cross at a kind of focus when a cup-shaped kathode is employed. The same may be distinctly shown by using a hemi-cylinder as kathode, such as *a* in Fig. 50, connected by the wire *b* with the negative pole N. Opposite is the anode P. The projection of the molecules in this case to a *linear* "focus" will be beautifully shown by the phosphorescent pattern produced by the thence diverging streams upon the surface of the glass.

74. Molecular Shadows.—Further interesting phenomena must follow, if it be a fact that radiant matter does not permeate the whole tube, like the brush discharge in a low vacuum, but comes from the kathode in normal and straight lines. Obviously if such straight radiating lines of discharged molecules can be obstructed by solid matter, and so prevented from striking the glass, a sort of molecular *shadow* must be produced. This we find to be the case. Let us take a somewhat pear-shaped tube, as in Fig. 51, having the negative pole *a* near the narrow end. Rather beyond the middle is a cross *b* of thin aluminium or mica, which (for convenience only, as it has been already explained that it matters little where the anode really is) is connected to the positive pole P. Where the molecular streams projected from the kathode *a* are stopped by the cross, there is a non-phosphorescent shadow *cd* visible on the large end of the tube, the rest being brilliantly phosphorescent.

FIG. 50.—Hemi-Cylindrical Kathode.

75. Molecular Effect of Bombardment.—If the

phosphorescence is produced, as supposed, by actual molecular vibrations set up by this energetic bombardment, we should expect to find more or less permanent signs of such mechanical action; much more supposing, as is very probable, that there may be also a giving up by the molecules of gas of their negative electrical charges. And there is abundant evidence of this. Some permanent molecular effect of long continued bombardment has already been noticed, in the gradually changed colour of powdered alumina; but there are others which are manifest in a much shorter time. Perhaps

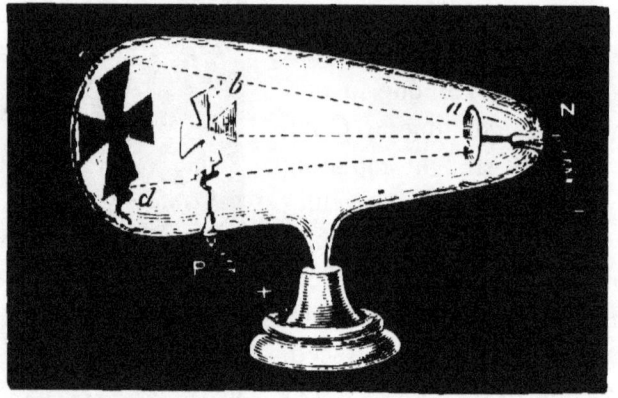

Fig. 51.—Molecular Shadow.

the first to be observed is a kind of molecular *fatigue*, which after a while deadens or lessens the luminous effect of the bombardment. This is easily shown. Let the cross *b* in the last figure be mounted on a hinge, so that after the shadow effect has been maintained for a certain time it can be shaken down, out of the way. Previous to this the shadow has appeared dark on the end of the tube, as at *c d* in Fig. 52; but directly the cross is removed it appears *brighter* than the rest, as at *e f*. The part outside the shadow is really just as bright now as the moment before; but it has been gradually becoming less brilliant than at first, and so the protected and as yet unfatigued shadow of the cross now appears more brilliant in comparison. Of course this difference would, from the same cause, now gradually disappear.

This, however, is not all the effect of the bombardment. After a certain time the glass of the tube more or less recovers its power of phosphorescence. But it never does so fully, or to the original degree; hence this form of tube should be sparingly used. It is remarkable that the bombardment should thus obviously produce some permanent change in the glass. Mr. Crookes found that when a tube had been worked until a very strong "phantom," or negative shadow, had been thus produced on the end, even when the glass

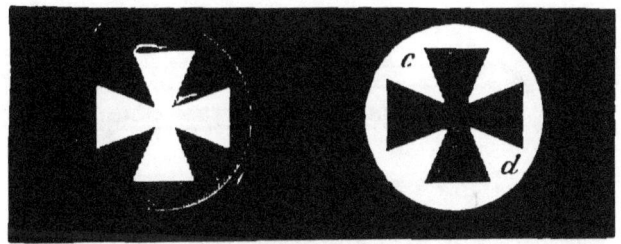

FIG. 52.—Molecular Fatigue.

was afterwards heated to the melting point, and then restored to its former shape and re-exhausted—even then the molecular difference shown by the phantom shadow had not disappeared.

It cannot be said to be positively known whether this effect of bombardment is chiefly mechanical; or owing to the driving of molecules of the gas into the glass; or due to the rapid communication of so many molecular electrical charges as must take place. Further experiments by Gouy proved that when a Crookes tube which had been exposed to energetic bombardment was heated, a "matt" surface was developed; and, if the heating was pushed to the point of actual fusion, minute but distinct bubbles of gas became visible. This phenomenon rather points to the effect being due to occlusion of gas. The fact that in using a tube for work with Röntgen rays the vacuum gradually increases, (§ 90, 91) also seems to point to the conclusion that gaseous molecules do actually penetrate the glass by the force of their impact.

76. **Mechanical Effects.**—There can be no doubt that the chief mechanical effect of these molecular impacts is distributed among the *molecules* of the bombarded body, as shown in heat or phosphorescence; but if the body be light and very easily moved, such mechanical impacts ought also to reveal themselves in movement of its entire mass. Mr. Crookes' apparatus for demonstrating this is shown in Fig. 53.

FIG. 53.—Mechanical Effect.

The exhausted tube has a little glass railway running along its entire length, on which rolls the axle of a light wheel carrying mica paddles. At each end of the tube, on a level with the upper paddles, is a pole in the shape of a flat aluminium disk. Whichever pole is made the negative or kathode, from it the radiant molecules of the exhausted gas are propelled along the tube and, striking the upper vanes of the paddle-wheel, cause it to revolve rapidly and travel along the rails. Then reversing the connections, the wheel will be seen to travel back again; and, if the tube be inclined, it will be found that the molecular impacts have force sufficient to drive the wheel up a quite perceptible slope or hill.

Another experiment brings out clearly the molecular nature of these streams from the negative pole. The tube shown in Fig. 54 has a central stem *a* topped by a needle-point, on which revolves very freely the glass axis of a light fly-wheel *b b*, whose vanes are of thin mica set at an angle of 45° to the horizon. Below the wheel, and under the centres of its vanes, is a platinum ring *c c*, wires from the ends of which are sealed into the tube at *d d*. The positive pole is sealed into the tube at *e*. When the negative terminal of the coil is connected with *one* only of the wires at *d* and the positive with *e*, owing to the stream of negatively electrified molecules from the ring, the vanes (which are bombarded diagonally by them) rotate with great rapidity. So far there is nothing more than in the preceding experiment. We may now disconnect the coil altogether, and connect instead the *two* wires *d d* with a galvanic battery, the effect of which is simply to *heat* the ring *c c* to the point of redness. The vanes now rotate as fast as they did with the coil; and we know that the chief effect of heat is to increase the molecular motion in a gas.

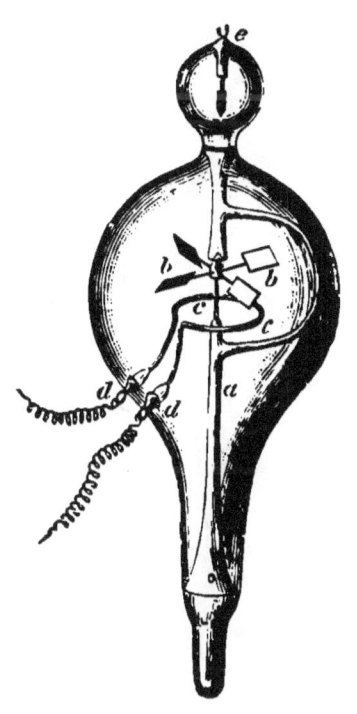

FIG. 54.—Discharge of Heat.

77. Recoil.—Action involves reaction; and any mechanical effect really due to propulsion of molecules from the kathode, must have a reciprocal in recoil of the kathode itself, which should be made manifest if the kathode be made the movable part of the arrangement. This is clearly exhibited by the apparatus shown in Fig. 55. The negative terminal N is here connected with a steel point, on which revolves an aluminium fly-wheel, which thus becomes itself the kathode. The positive pole P is sealed into the top of the tube. To confine the molecular radiation, or at least to make it preponderant on one side of the vanes, one side of each vane is covered with thin mica. On passing the discharge the vane rotates rapidly. The exhaustion in this case need not be so great as in some other experiments; it is sufficient if it be a little more than enough to make the dark space on the metal side of each vane extend to the side of the tube.

78. Effect of a Magnet on Radiant Matter.—Though there is a difference from the phenomena as seen in low vacua, which will appear presently, electrified molecules in motion in a high vacuum must be also attracted by the poles of a magnet. Fig. 56 shows a long tube whose negative terminal N has a disk kathode a, in front of which is a mica screen $b\,d$ with a small hole e in the middle; the positive pole is at P. The result of this, when the current is passed, is a molecular discharge along the central line $e\,f$, which is readily made visible by a longitudinal phosphorescent screen $b\,c$ along the middle of the tube, on which the line of discharge shines out clearly. Bringing under the tube a strong compound magnet, this line of light is found to be drawn down to $e\,g$, and can be made to wave about curiously as the magnet is moved to various positions.

The amount of such a deflection must vary with the proportion of the magnetic force to the resistance to the motion,

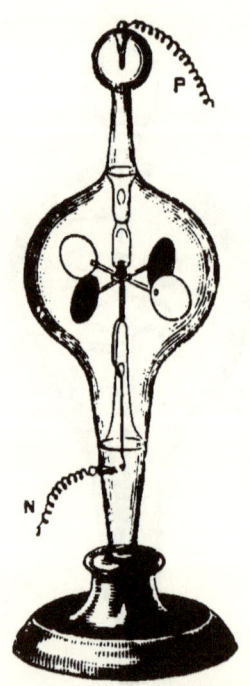

FIG. 55.—Recoil.

which we may fairly assume to vary with the degree of exhaustion. This may be demonstrated by a tube prepared exactly as the last, but with the addition of a small potash-tube

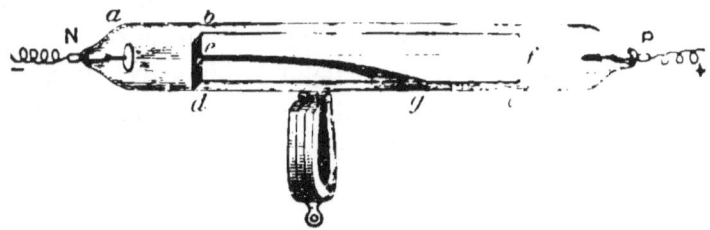

FIG. 56.—Magnetic Deflection of Discharge.

at one end, by heating which we supply vapour to the vacuum, exactly as in the experiment described in § 72. With the tube in its exhausted state the discharge takes the longer trajectory. On heating the potash, and so diminishing the vacuum, it is just as if the path of a number of projectiles were more obstructed by a denser medium; and the discharge

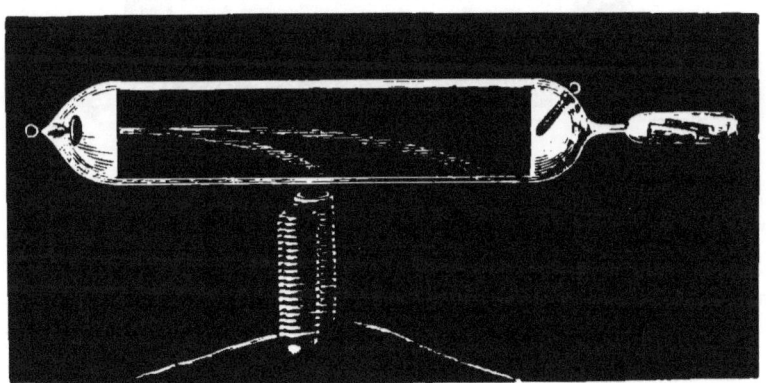

FIG. 57.—Vacuum and Deflection.

is further dragged down as in the shorter curve, the gradual change in the curve being clearly shown on the phosphorescent screen.

The magnetic deflection can also be made visible mechanically, and in a form, too, which by an optical lantern can be readily shown upon a screen.[1] Fig. 58 shows the apparatus. The negative terminal N has for kathode a shallow cup *a b*, some way in front of which is fixed a mica screen *c d* large enough to intercept the conical stream of molecules proceeding from the kathode. Behind this screen is a mica paddle-wheel *e f,* and the positive pole is at P. With this arrangement alone the wheel would not turn, being shielded

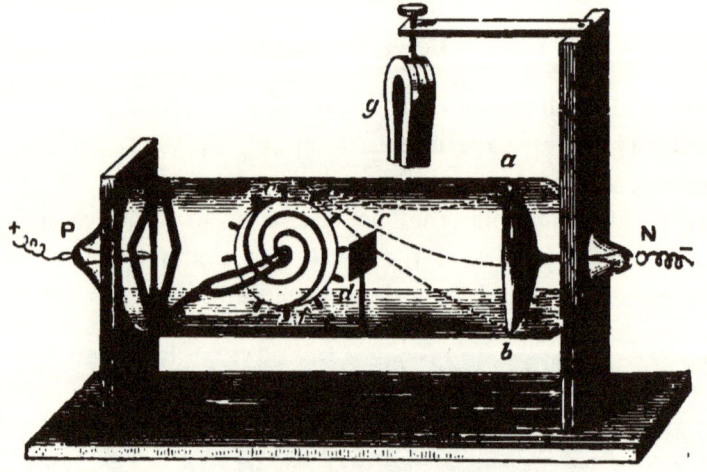

FIG. 58.—Deflected Discharge.

by the mica screen. But a strong magnet *g* being placed over the tube, according to the way the poles are turned, the molecular cone will be deflected upwards (as shown by the lines) or downwards, and the wheel is caused to spin rapidly. The motion is in fact generally too rapid to be seen, but for the spiral construction of the body of the wheel, which enables the direction of the motion to be readily shown by the apparent expansion or contraction of the spiral.

[1] It is not possible here to enter into details as to the screen projection of such experiments. Full information on such subjects will be found in *Optical Projection* (Longmans) by the present writer. A good example will however be found in Fig. 65 a little further on.

79. **Difference in High and Low Vacua.**—We know that the discharge in an ordinary vacuum-tube is also deflected by a magnet; but there is a very interesting difference. In the foregoing experiment the deflection becomes *permanent* in the altered direction caused by the magnetic force—there is no recovery. It is quite different in the low vacuum. Let us take an ordinary vacuum-tube (Fig. 59) with poles at P and N, and pass the discharge through it; the exhaustion should be such that the current passes as a line of violet light between the two poles. Underneath arrange a powerful electro-magnet. On passing a current through the magnet, this discharge also is deflected, up or down, according to the direction of the current

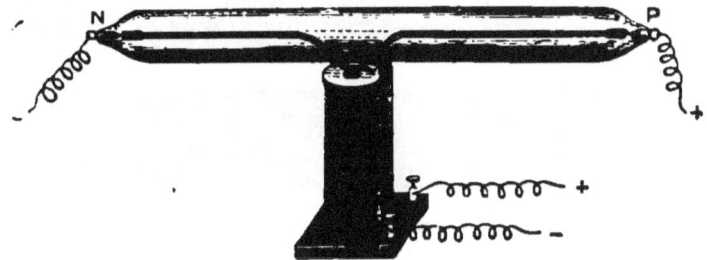

FIG. 59.—Deflection in Low Vacuum.

and consequent position of its poles; here it is shown deflected downwards. But the dip in the line of discharge is only temporary; after that it rises again, and pursues its former direction.

It appears that in this last case we have to do with a *flexible connecting current* proceeding from the positive pole, while in the high vacuum all the phenomena unite in pointing us to a stream of negatively electrified molecules *repelled* with great velocity from the negative pole. This is further made clear by the arrangement shown in Fig. 60, where the highly exhausted tube has two negative terminals a and b, and one positive at c. In front of the double kathode is a mica screen with two apertures $d\ e$, and up the middle of the tube is a phosphorescent screen, by which is clearly shown the path of each discharge. We know very well that two nearly parallel *currents* in the same direction attract each other, and if allowed will approach, as do two movable wires carrying currents; and this can easily be also verified as a fact with two currents in low-vacuum tubes. On the other hand, similarly *electrified particles* repel each other. The experimental tube here shown is so constructed that when the negative discharge is only passed from *a d*, its path is *d f*, and when from the kathode *b* only, its path is *e f*. But when the coil is connected by a

Fig. 60.—Repulsion of the Molecules.

branched wire with both kathodes, the two lines of discharge are at once *repelled* from one another into the more diverging lines *d g* and *e h*.

80. **Magnetic Rotation.**—This deflection of the discharge in the magnetic field is due to the *torque* or twisting effects due to each of the two poles combined. Fig. 61 illustrates the nature of the effect. The north pole gives the discharge of molecules a twist one way, and the south pole the other way; the two poles together compel a deflection up or down, in a plane at right angles to the line joining the poles.

If we enclose the tube itself in one axis alone of a strong magnetic field, as we may do by enclosing it in a coil traversed by a powerful current, we get the torque or twist alone. Using thus a tube resembling Fig. 51, but arranged with rather a longer distance between the cross and the phosphorescent end of the tube, the molecular shadow of the cross is more or less sensibly *rotated* on the end of the tube. It is as if a polarised optical image was rotated by similar means. The ordinary straight streams from the negative pole under these circumstances become twisted.

So much had been well known, and indeed was predicted before it was observed; but quite recently[1] Prof. J. A. Fleming has discovered further very interesting phenomena. He surrounded the tube axially with a coil of about a hundred turns

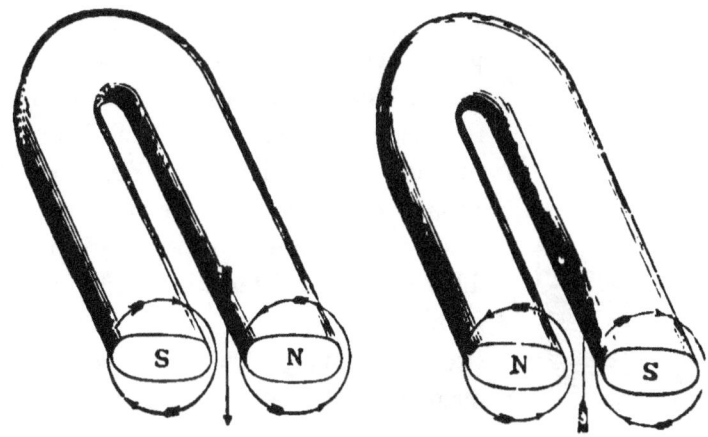

FIG. 61.—Nature of the Deflection.

of wire, through which a current could be passed ranging up to 10 or 12 amperes; the coil being placed between the cross and the fluorescent end of the tube. As the current was gradually increased the shadow of course was more and more twisted; but at the same time it became gradually smaller and smaller, and the arms became curved instead of straight. The fluorescent spot also decreased in size. Finally the cross and original fluorescence disappeared, a very small shadow being however visible as long as any bright green was left. Still increasing the current, a new and larger shadow of the cross

[1] See *Electrician*, Jan. 1, 1897.

appeared, also slightly twisted, and not so defined in outline, which also diminished in size with further increase of current. This sequence of phenomena appears to show a progressively greater convergence of the parallel molecular streams from the disk kathode, until finally they again *diverged* as from a point to form the larger shadow. On adjusting the coil of wire between the kathode and the cross, the shadow was found to be diminished, but not screwed or distorted.

Sir David Salomans found that if the single pole of a strong bar-magnet kept parellel to the axis of the tube, was presented close to one arm of the cross, and the magnet moved round (still keeping it parellel, so that its length described the surface of a cylinder) the shadow of the cross either followed or preceded the pole. This is a most interesting and pretty experiment, very easily made either with a strong steel magnet, or a straight electro-magnet excited by a small independent current.

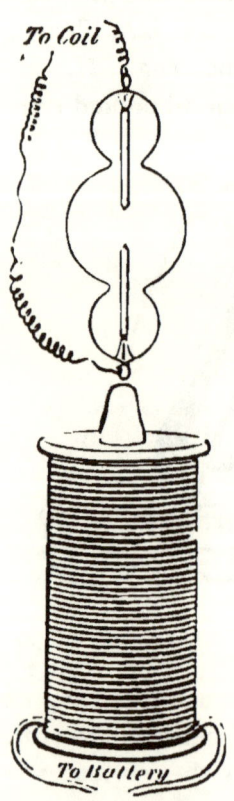

Fig. 62.—Crookes' Tube for showing Magnetic Rotation.

Mr. Crookes made many attempts to produce actual visible rotation of the discharge, analogous to De la Rive's experiments in low vacua (Fig. 27). He succeeded at length by the arrangement shown in Fig. 62. A bulb was blown of German glass, with a smaller bulb at each end communicating with it by a short neck. At each end is a long aluminium pole, turned to a cone at the end. The discharge fills the centre bulb with a fine phosphorescent green light, while the neck surrounding the negative pole exhibits two or three dark and bright patches in constant motion, which follow one another round, first one way and then reversing, also every now and then dividing or fusing into one another. Owing probably to the magnetism of the earth, after a while the motion often becomes more regular and slow, the patches in the neck generally resolving into two or three, and the green light in the centre bulb becoming more intense along two opposite lines parellel to the line joining the poles, forming two faintly outlined brighter green patches, which slowly move round the bulb a semi-circle apart. Bringing a strong electro-magnet underneath, as shown in the figure, these more or less uncertain phenomena are converted into orderly rotation.

But we find still the difference between *molecular* and *current* discharge. In De la Rive's experiment, in low vacua, the rotation is reversed, not only if we reverse the magnetic field, but if we reverse the poles of the discharge in the tube; because it is a *current* which is concerned. But in Mr. Crookes' experiment in the high vacuum, the rotation is only reversed if we reverse the pole of the magnet; if we only reverse the current, as regards the two electrodes, it makes no difference.

FIG. 63.—Heat at the Molecular Focus.

81. **Heating Effects.**—The molecular discharge produces another very obvious effect, as we should expect. Wherever it produces phosphoresence on the glass the latter is warmed, and, if the bombardment be brought to anything like a focus, the heat is intense. Mr. Crookes demonstrated this by two striking experiments. Fig. 63 shows a bulb whose negative electrode

N has a cup-shaped kathode *a*, at whose centre of curvature is placed a piece of refractory metal *b*, usually either platinum, or the still more infusible alloy of platinum and iridium. Using a very moderate current, this is readily brought to a white heat; and if the vacuum be properly adjusted, the full power of a good coil will actually melt even the iridium compound, and still more easily a piece of platinum alone.

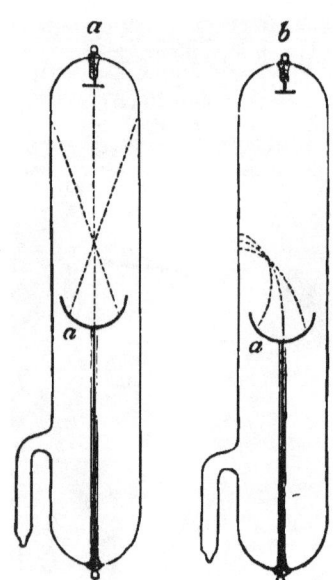

FIG. 64.—Perforation of a Tube.

Or the heating effect can be shown in the perforation of the tube itself. Fig. 64 shows a narrower tube with the cup-shaped kathode *a*, which in the diagram *a* brings the molecular streams to a focus in the centre of the tube, from which they diverge to render the positive end phosphorescent. This diverging cone only makes the tube warm. But by using a magnet, as in some preceding experiments, the cone can be so deflected to the side of the tube (as in diagram *b*) as to strike there nearly in focus; the effect is then so great as to melt the glass, and by the pressure of the atmosphere the tube is perforated. Fig. 65 shows the arrangements by which the perforation of a tube by this effect of the discharge was shown to the entire audience by projection upon a screen, before the Royal Institution, and again before the British Association in 1879.[1] At *d* is the lantern, with its own current wires, and in the foreground the battery and coil, the latter connected up to the tube *a b*, supported in the field of the lantern, *b* being the cup-shaped kathode. The tube is then

[1] The electric lantern was employed in both cases; but it may be worth observing that an oxy-hydrogen lantern is amply sufficient.

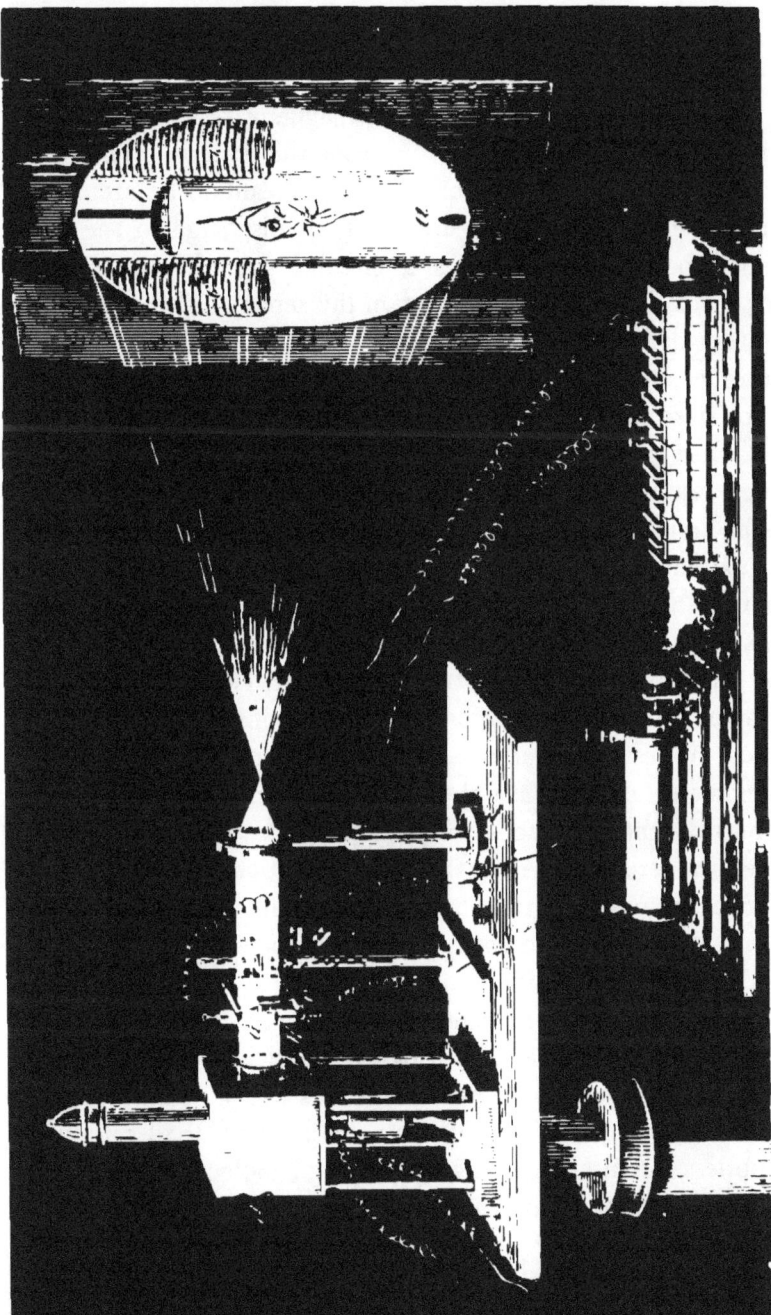

focused on the screen in the enlarged image *a b*. In front of the tube is adjusted the electro-magnet *c*, to which another pair of wires can be switched on from the battery. The tube is thinly coated with wax on the side meant to be attacked. The first effect of deflecting the cone of discharge to the side of the tube is the melting of a circular patch in the coat of wax, which is clearly pictured upon the screen. Then the glass is seen disintegrating, with small cracks diverging from the centre of heat; finally the glass melts, the pressure of the atmosphere forces it in, and a perforation takes place as shown at *e*, when the experiment comes to an end.

82. **Nature of the Discharge.**—Such is a brief outline of the series of discoveries and experiments which established the existence in high vacua of a peculiar kind of discharge from the negative pole or kathode. This Mr. Crookes held to consist of molecules of the residual gas, negatively electrified, and then driven by electrical repulsion violently from the kathode. In some respects these streams of molecules seemed almost to resemble radiant energy, and most of the ordinary phenomena of gases are absent in this state of matter; yet, as regards chemical action and affinity, the radiant molecules were as material as ever, so that the exhaustion can be increased beyond the power of the pump alone, by the use of absorbents in the usual way. The earth thorina was found to be a specially good absorbent of residual air.

Continental men of science, however, very seldom seem to accept readily the views of English physicists, and Crookes' view as to the nature of the discharge was by no means accepted by them. Puluj for a long time maintained that, although doubtless the discharge was a stream of molecules, they were molecules of metal torn off from the kathode. Such particles *are* torn off, and after a while are deposited as a film on the inside of the tube. So far as this process extends, it is perhaps even possible that such molecules may take some little share in the phenomena. But that it can only be a proportionate and very small share, not at all of the essence of the phenomena, Mr. Crookes finally proved by the simple means of repeating all the essential experiments, including that of the mica cross shadow (Fig. 51) and the rotating mill (Fig. 53), with tubes containing no inside metal terminals whatever, but using external electrodes of tin-foil fastened to the *outside* surface of the tube. Rubies were also rendered brilliantly phosphorescent with such external electrodes. It has also been shown that under these conditions the radiant particles are not torn off from the glass, at least not to any appreciable extent.

Wiedemann, Hertz, and the majority of German physicists,

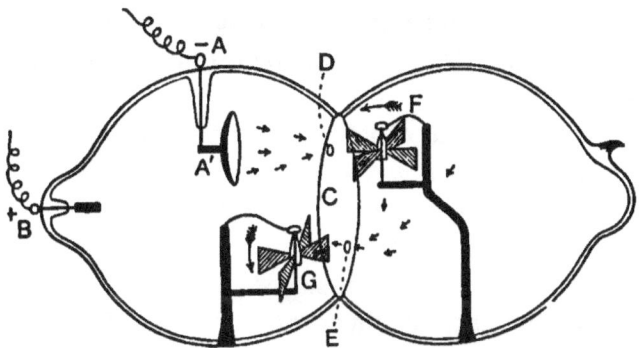

FIG. 66.—Return of Gas Molecules.

on the other hand, adopted the idea that the "kathode rays," as they called them, were not streams of particles, but some kind of wave-motion, analogous to light, produced in the ether. It is difficult to see how this view can be reconciled with some of the experiments already described, especially the visible repulsion of two streams of discharge (Fig. 60) and the deflection by a magnet (Figs. 56, 57, 58). For these latter must obviously imply some organised *structure* in the ether, of which no other signs are known to exist, if ethereal waves are the essential part of the phenomena. But as further proof of his own theory, Mr. Crookes has since devised the ingenious apparatus shown in Fig. 66. It is pretty manifest that if molecules of gas are continually repelled from the negative pole, unless the supply of molecules finds its way back again, the phenomena must ultimately come to an end. In this apparatus the rarefied tube is therefore divided in two by a glass screen C, pierced with holes at D and E. The hole D is in the focus of the concave kathode A' connected with the negative wire A, and behind it is adjusted the fly F, so that its vanes come successively in the way of the stream through D. The fly rotates when the discharge is passed, as we have already seen. Opposite the other hole E is another fly G. This fly also rotates, showing thereby that a return stream of molecules is passing back in the reverse direction through the hole E. This fact, it will be seen (§ 92), seems to be of considerable importance in the design of the best tubes for use in Radiography.

Another remarkable experiment tending the same way was made by Holtz. He found that when a cylindrical tube was furnished inside with inclined annular glass obstructions, in the form of funnels (or truncated cones, the central aperture taking the place of a spout) the resistance was much less when the concavity of the internal cones or funnels faced the negative pole. If the discharge is other than material, it is difficult to account for this.

Finally, in 1893, Prof. J. J. Thomson made experiments to determine the velocity of the discharge by means of a rotating mirror.[1] It was found much smaller than that of light, or of the ordinary discharge in an ordinary low-vacuum tube, and approximately the same which negatively-electrified atoms of hydrogen would acquire at the determined potential of the kathode.

83. Effect in Vacuo.—By the use of absorbents, together with persistent heating and pumping, tubes can be prepared with residuals not probably exceeding 1-twenty-millonth of an atmosphere. This is not a perfect vacuum, and consequently

---

[1] See *Phil. Mag.* xxxviii. 358.

not a perfect di-electric; but it is sufficiently near the limit to prevent the spark passing any appreciable distance. If the electrodes are brought even to within 1 mm. of each other, the discharge will pass as a "spark" through the external air, between the external connections, rather than traverse the short distance within the tube.

84. **A Paradoxical Experiment.**—We may conclude this chapter with an apparently paradoxical experiment devised by Hittorf[1] to demonstrate that when the distance between the electrodes is *less* than the "dark space" around the kathode,

FIG 67.—Hittorf's Experiment.

it is very difficult indeed for the spark to pass; while, if the distance be increased, it may pass readily. He arranged a tube, as in Fig. 67, the main portion consisting of two bulbs connected by a tube of small diameter, in which the two electrodes approached within a distance of 1 mm. The bulbs were otherwise connected by the spiral tube, whose total length exceeded 1 metre. When the vacuum was made high enough, the discharge all passed in preference by the long spiral tube!

In this case there can be no doubt that the discharge is assisted by the walls of the narrow tube. But if a bulbous

[1] *Wied. Ann.* 1884, p. 96.

tube is furnished with convex electrodes, and gradually exhausted, it will be found that after a certain point the discharge leaves the central portion of the electrodes, where the distance is least, to retreat more and more towards the more distant portions of the curved surfaces, thus taking by preference the longer distance.

The fact thus demonstrated also has an important bearing (see § 92) upon the construction of tubes for producing the most powerful Röntgen X rays.

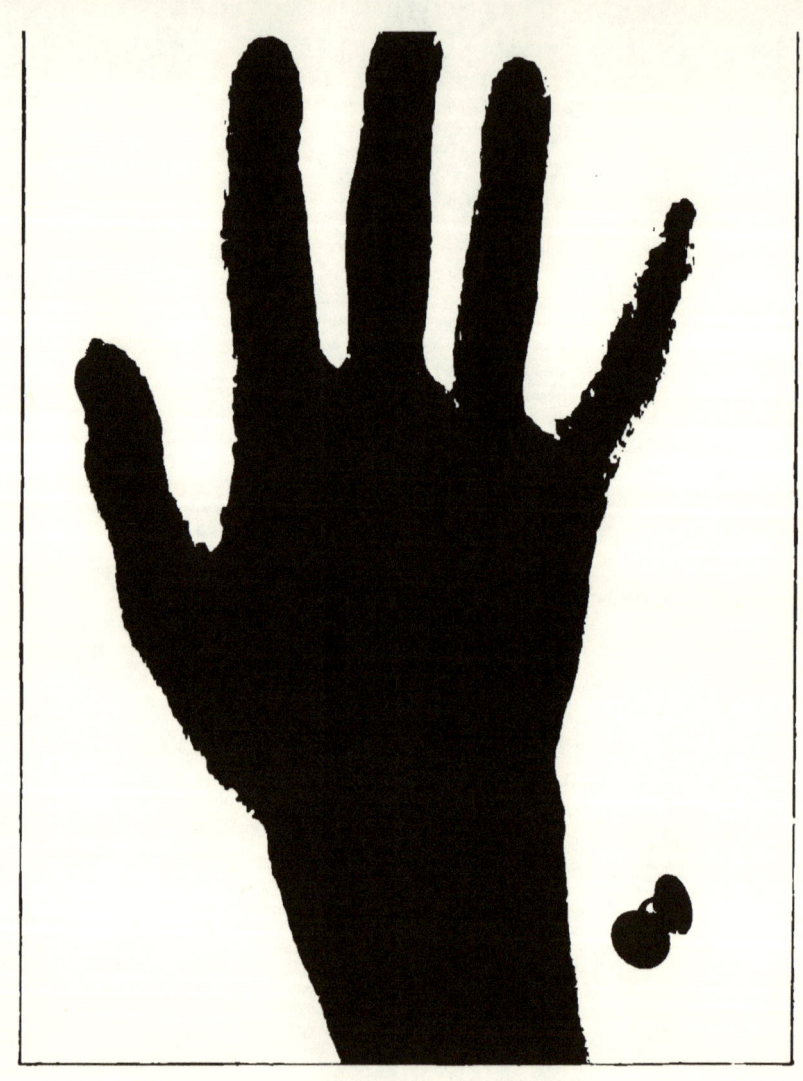

PLATE I.—HAND.
Showing generally shadows of Bones, Flesh, Sleeve-links, and faintly of linen Wrist-band.

CHAPTER VIII

RÖNTGEN X RAYS

A NEW field of work and of investigation was opened by the discovery in 1895 of these rays by Professor Röntgen of Wurzburg.

85. Kathode Rays.—It has been already mentioned (p. 82) that while English physicists generally adopted the view of Mr. Crookes, and considered the radiation from the kathode in a highly exhausted tube to consist of streams of negatively-electrified molecules, Hertz and many leading Continental physicists clung to the theory that it was of the nature of ethereal wave-motion. They accordingly gave to the discharge the name of "kathode rays." Whatever they are, we have seen that mica acts as a screen to them, and the glass of the tube also stops them. But Hittorf discovered that they appeared to pass through a thin plate of aluminium within the tube; and in 1894 Professor P. Lenard made a further remarkable discovery. Employing a tube something like Fig. 68, in which the end facing the kathode was closed by a plate of aluminium, he found that *something* radiated from the *other* side of the aluminium plate, which excited fluorescence in a fluorescent screen, and produced photographic effects upon a plate, even through additional plates of aluminium or of quartz. He further found that these rays—whatever they were—were deflected by a magnet, which altered the position of the fluorescent spot upon the screen. These outside rays were thus more or less identified with the well-known kathode rays within the tube, though we now know that they were in reality copiously mixed with true Röntgen rays. It appears, in fact, that Professor Lenard was only prevented by pre-occupation with the ethereal hypothesis concerning kathode rays, from making the further discoveries of Professor Röntgen. The most natural view would indeed appear to be, that it must be something of the character of ethereal motion which traversed the aluminium plate; but this portion of the phenomena (*i.e.*, outside the tube) is now generally accounted for by English physicists on the supposition that the bombardment of molecules on the

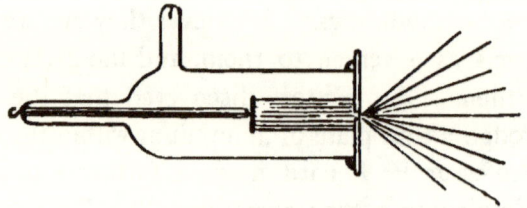

Fig. 68.—Lenard's Experiment.

inside of the plate, causes a corresponding discharge or repulsion of other molecules (not the same) on the outer side.

86. **Röntgen's Discovery.**—In November, 1895, Professor Röntgen discovered that some kind of radiant energy, emitted from some part of a Crookes' tube he was using, affected a photographic plate near it, which was quite shut up or enclosed in the usual "dark back," or case of wood. He discovered that these rays passed freely through black paper, cork, wood, &c., and also through his hands, but less freely through the bones than the flesh. By laying his hand upon a plate enclosed in thin wood or black paper, with an excited Crookes' tube a little above the hand, a photograph clearly showing all the bones was obtained. Further investigation taught him that these rays also excited powerful fluorescence in certain substances; so that in a darkened room the bones of the hand could be seen depicted upon a screen coated with barium

platino-cyanide. He yet further found that they appeared incapable of refraction, regular reflection, or polarisation; and also that they were *incapable* of deflection by a magnet; this last feature clearly distinguishing them from the already known "kathode" rays, and also from Lenard's rays. The uncertainty as to their actual character, which these puzzling facts presented, led Professor Röntgen to give these new rays the provisional and singularly appropriate name of "X" rays. He himself, at that time, believed them to consist of *longitudinal* vibrations in the ether, set up by the bombardment of the molecules upon the wall of the tube.

87. **Nature of the " X " Rays.**—A host of physicists attacked the new problems thus presented; but it is remarkable how very little that is fundamental has been added to Röntgen's original discoveries. It was soon found that transparency to these rays diminished, roughly speaking, very much in proportion to molecular weight. Thus, aluminium 15 mm. thick was about as transparent to them as crown glass; but flint or lead glass is very opaque, and lead 1.5 mm. thick is practically opaque. Powders are generally as transparent as solids! These facts concerning transparency led Boltzmann to suggest that "kathode" rays were longitudinal short waves in the ether, and hence rapidly absorbed in the air; and that the X rays were, on the other hand, longitudinal vibrations of comparatively great length, but high frequency; the length explaining the transparency, and the rapidity the fluorescence and photographic effects.

That view is now generally given up. Becquerel discovered that uranium and its salts, and some other fluorescent substances, after exposure to light, also emitted rays which passed through bodies opaque to visible light, and would give photographic images; such fluorescent substances also continued to emit these rays even for weeks in the dark, the amount of the invisible radiation bearing no apparent proportion to ordinary fluorescent rays from the substance. These invisible rays, however, were found to show distinct traces of polarisation and refraction, though much less than ordinary light, or even than the well-known "ultra-violet" rays of the spectrum. In most respects they are, in fact, pretty obviously intermediate between ultra-violet and X rays. It was even found that Becquerel's rays gradually discharged an electrified body; or, what is the same thing, transformed the air through which they passed into a conductor. This remarkable effect is produced still more strongly by the X rays. It may be demonstrated in many ways, but one of the simplest and most elegant I owe to Dr. J. Mackenzie Davidson, of Aberdeen. The outer coat of a Leyden jar is connected to earth, and the jar charged. Near the knob is then suspended by a silk fibre a very small pith ball surrounded by a little crumpled gold-leaf. This will be repelled as usual. Turning on a powerful discharge of the rays, the ball will gradually swing back, to be again repelled. The effect is easily projected upon a screen.

The Becquerel rays were, however, obviously invisible light-rays of an extremely ultra-violet character; *i.e.*, transverse vibrations in the ether of extremely short period and wave-length. It is now generally held that Röntgen's X rays are ether-waves of the same character, but still higher frequency—certainly not less and possibly much more than a hundred times that of visible green light. The absence of sensible refraction, polarisation, and regular reflection is exactly what would then occur. The waves are probably simply too small to be reflected from any polish at present attainable, or polarised by any ordinary structure in sensible degree. The wires of a bird-cage are found to plane-polarise the long Hertzian electric waves; and wires strained close together were found by Imbert and Bertin-Sans to polarise the infra-red rays of the spectrum, but fail with shorter waves. Just so, the X waves are believed to be too "fine" for the structure of a tourmaline. It is moreover believed that later experiments by Prince Galitzine and Karnojitsky, and some other physicists, do show such faint traces of regular reflection and polarisation, as could alone be expected upon such a hypothesis. Sir G. C. Stokes further believes that these extremely short transverse waves are *non-periodic;* but on this point there is much more doubt, and some recent experiments appear to show traces of interference, which make it more doubtful still.[1]

88. Practical Applications.—The practical applications of these phenomena were at once seen to be numerous and important. In merely physical matters, real gems are found to be almost transparent to these rays; while paste imitations, of dense lead glass, are almost opaque. The utility of this is obvious. Rücker, again, found some specimens of porcelain almost as transparent as glass, while others, in which bone-ash was largely present, were nearly opaque—in certain cases this is highly useful to a collector. Cracks and flaws and other variations in metals have been made apparent by these rays; and it is impossible to foresee to what extent this kind of practical application to merely physical purposes may be carried.

But (as was indeed foreseen at once) it is to physicians and surgeons that Röntgen rays have proved of the most inestimable value. They were found so within two months of publication of their discovery; and before six months had passed they were employed almost daily in the great hospitals of Europe. By their aid were, and are, now easily detected needles and other foreign substances in the body, malformation of the bones, ossifications, calculi, badly-united fractures, and a

---

[1] The latest investigations still leave the matter uncertain. Dr. Fromm, of Munich, found what he considers interference effects, which give broadly about $\frac{1}{78}$ of the shortest visible violet rays. On the other hand, Dr. Precht, of Heidelberg, finds an exceedingly complex radiation, of which he thinks a portion is not wave-motion at all, and that the Röntgen waves are too long to pass through black paper as transverse vibrations. Hence he still inclines to the idea of some at least being longitudinal waves. Proof of the latter would be of profound interest, but on the whole the theory of very short and rapid transverse vibrations still holds the field.

host of pathological conditions. Some of these are best photographed; in other cases the shadow upon a fluorescent screen answers all purposes. As better tubes and more powerful currents came into use, the penetrative power of the rays increased; and photographs were taken or fluorescent images observed, through the entire adult human body; by the latter method, the beating of the heart and motion of the diaphragm are readily made visible upon the screen. Success in these applications, and greater progress in this direction, will depend upon the understanding and fuller investigation of those different conditions of the rays to be presently indicated; but there is no doubt that—to take one well-remembered case—the life of General Garfield, formerly President of the United States, would have been saved if his surgeons had had the assistance of the Röntgen rays. Where probing is useless or unsuccessful, because a bullet or other foreign body has lodged behind a bone and cannot be felt by the probe, its precise locality is now indicated; or where diagnosis is doubtful between mere fracture or mere dislocation or a complication of both, the state of affairs can be at once known; or in case of a "stiff" joint, these rays reveal whether it is a growth of bone or of softer tissues which prevents the movement.

Other applications have been made by physiologists. At an early period, Dutto injected the arteries of a hand with a solution of calcium sulphate, which is tolerably opaque to X rays, and thus obtained a really beautiful photograph of the entire arterial system. And countless curious variations in the osseous system have been already discovered and photographed.

89. **Source of the Rays. Tubes.**—We come now to practical work and details. As these rays are not sensibly refracted, photographic or other images are necessarily taken as simple, direct *shadows* cast by the objects. The best and smallest source of radiation is therefore a matter of the greatest importance. The source of radiation in the first instance was

found to be the phosphorescent surface of the Crookes' tubes. It soon became manifest that the source was, broadly speaking, the *first solid surface* struck by the kathode rays, or bombarding molecules. Salvioni and others obtained sharper shadows by using a concave kathode, and directing the "focus" to some point on the wall of the tube by a metallic rod held in the hand, which could shift the focus to another point when the first was used up or "fatigued," as explained in the last chapter. But while these experiments were being made, Mr. Herbert Jackson, of King's College, during a series of investiga-

Fig. 69.—Focus Tube.

tions of his own in phosphorescence, had already determined the best form of tube ; one which still holds its ground, and, under the name of the "focus tube," is used everywhere in work with these rays. It is a modification of the Crookes' tube already given in Fig. 63, and is shown in Fig. 69. The ordinary anode is replaced by a flat plate of platinum, placed at an approximate angle of 40° with the axis of the tube, in front of a concave kathode of aluminium, which should be as *smooth* on the concave surface and edge as possible. The kathode need not "focus" actually on the platinum plate ; on the contrary the latter, which we will call the radiating plate, or radiator, is better somewhat *beyond* the focal point, the exact distance being immaterial within moderate limits so far as focus or definition is concerned,[1] as the kathode rays, after converging to a focus, do not appear to diverge, at least when the vacuum is at a proper degree for efficient work (see p. 134). None of the modifications since described in this tube have at all changed it in these essential characters; and it is doubtful if any of the supposed improvements have proved of advantage, except perhaps the employment occasionally of a double anode as presently described.

90. **Vacuum, Current, and Character of Rays.**—It was soon discovered that unmistakable Röntgen rays were considerably different in character; and the more rapid work with "focus" tubes brought these differences into clearer light. It is important to understand them. Battelli and Garbasso seem to have been among the first to observe that ordinary Crookes' tubes generally gave at first feeble results, which improved during seven or eight hours constant current, and diminished after a certain time. This they attributed to increase of vacuum from gradual occlusion of gas by the electrodes or walls of the tube, a conclusion which has been abundantly verified.

Professor S. P. Thompson investigated the matter more in detail. Such experiments are best made with a fluorescent screen to show the results. During steady exhaustion of a tube he found the following order of phenomena. (*a*) With still a low vacuum and electrical resistance small, so that sparks of 2 or 3 mm. would scarcely pass in preference between the dischargers, the earliest phosphorescent gleams appear on the wall of the tube, but none on the screen : a photographic plate will, however, give an image of easy subjects with a long exposure. (*b*) As exhaustion is carried on, there is an almost

---

[1] As noted in § 92, when the radiator is used as the only anode, the nearer it is to the kathode, the greater is the amount of resistance, and consequently the current used.

sudden increase in resistance, the discharger-spark rising to several cmm. in length; and at the same moment a strong fluorescence appears on the screen. The transition is less sudden with a new tube than with one already used, re-filled, and then re-exhausted. Two other points are to be noticed at this stage: the whole tube, both before and behind the inclined anode, fluoresces, except in a line in the plane of the anode; and these rays do not show nearly so much *difference* between the bones and flesh of the hand as at the next stage. (*c*) With yet further exhaustion, the discharger-spark of course lengthens; the part of the bulb in front of the anode is more brightly lit, while that behind is less so, till the tube behind is nearly dark, and the screen does not fluoresce at all to rays from the back of the anode. The *difference* of transparency between bones and flesh (of the hand) is now most marked, and, for work in this range, such is the best condition of the tube. (*d*) Carrying the vacuum still further, and necessarily pushing up the current to overcome the resistance, the rays penetrate both flesh and bone till there seems (in a hand) little difference of shadow between them.

Similarly, Chappuis and Nugues established by systematic investigation that with a given tube the maximum effect was obtained by a certain *frequency* of spark. With a certain tube, photographic effect fell off a great deal when great frequency was used; and as low as four sparks per second gave powerful effects, while also keeping the tube and electrodes cool. With another tube and a given coil, the maximum effect was obtained by ten breaks per second. It will be obvious that short and frequent spark is suitable for a low vacuum, while long and full sparks are needed for highly-exhausted tubes.

Mr. Herbert Jackson observed another interesting phenomenon during the course of the experiments which issued in his well-known "focus tube." As a high vacuum is approached, the molecular streams from the kathode are seen as faint blue rays meeting at its centre of curvature or focus, and thence diverging in another cone. As the vacuum increases, this latter or diverging cone narrows more and more, until at length it apparently shrinks into a mere straight line. So long as the tube is exhausted to this point, it does not seem to matter (in moderation) how far beyond the "focus" of the kathode the anode or reflecting plate is placed: the impact still remains sensibly confined to a point, and the shadows are sharp. This phenomenon is very peculiar and interesting; as physical and mathematical considerations appear to show that it is what would happen if we could conceive of the kathode rays as converging streams of molecules or atoms, which were *non-elastic* at the molecular distances concerned.

These variations in character and energy of the rays are of the highest practical importance.

91. **Management of the Tube.**—What has been said will make it very evident, that for a wide range of work, differing tubes will be required. Matters are, however, greatly facilitated by the fact that the vacuum steadily rises with hours of work, owing to occlusion of the gas molecules. Tubes are, therefore, usually supplied from the makers with a comparatively low vacuum, suitable for easy subjects; and by use they very gradually become suitable for longer sparks and more difficult operations. After long use, all tubes used with large coils finally become too rarefied for work of any kind.

This gradual hyper-exhaustion can, however, be controlled or counteracted to a very large extent; not only thus prolonging the life of the tube, but placing a very valuable means of control in the hands of the operator. Thus (*a*) by heating the tube with a Bunsen burner or spirit lamp, the vacuum is (potentially) lowered. (*b*) By sending a *much smaller* current[1] through the tube the reverse way for a short time occasionally, the vacuum is generally lowered. If the platinum radiator be the only anode, however, such a reversal deposits evaporated

---
[1] The full current reversed may probably wreck the tube.

platinum on the glass; and to avoid this some recommend a third terminal of aluminium, used only for such reversals. Most English workers, however, prefer heating. (*c*) It seems very obvious to apply Mr. Crookes' method of heating a small subsidiary potash bulb as in Fig. 47. In reality it is difficult, the vacuum generally lowering far too suddenly and too much, with the very least application of heat. This method appears, however, to have been too readily abandoned; for it seems likely that by using a *very small* particle of potash it might prove easy and successful. Professor Morton has used an auxiliary bulb, exhausted with the other, containing a carbon filament with two terminals—in fact, a small incandescent lamp; on passing a separate current through the filament a certain amount of occluded gas is given off. It should be worth while to use a very small portion of spongy platinum, and exhaust the tube with hydrogen gas. Several tubes use a small fragment of palladium, also exhausting with hydrogen.

An expedient, somewhat akin to heating by a burner, is to pass through a hyper-exhausted tube a very powerful current from a larger coil. Probably from the heating of the anode, which becomes red-hot under such treatment, the vacuum is lowered; and in most cases (which is somewhat difficult to account for, unless on the supposition that the platinum itself gives off occluded gas) this lowered vacuum has a considerable amount of permanence, and allows the tube to be again used with a coil of lower power than that employed in the renovation.

As a general rule, therefore, the chief means of control are (*a*) heating the tube (when required) by a spirit-lamp, Bunsen burner, or—what is often better—a common Argand burner turned down to a mere ring of blue spots; and (*b*) adjusting the distance between the dischargers. The proper effect is best judged in the first instance by the image on a fluorescent screen; and indeed, if it be screen work that is in hand, nothing further is required. In that case the heat is applied very easily, as the burner may be arranged under the tube, while the rays are reflected out horizontally. Having then got the tube into the condition required, or at least the best attainable condition at command, *switch off current* and set the discharger at a distance which allows sparks to pass, cutting the tube out. Now again *switch off current* and by a small degree increase that distance; and so on by degrees until the discharge *just* passes through the tube, in preference to the spark-gap. Then directly the vacuum increases, sparks will again pass, which gives notice that a little more heat is required. If the work is photographic, the lamp will generally have to be withdrawn to be out of the way, when the tube (if necessary) has been brought to the proper point as shown by the sparking. It will not always be found, however, that the best photographic result coincides with the best fluorescent result; as a rule the most powerful photographic result will be obtained with a slow, full, long spark at the interrupter, or with a mercury interrupter, which would cause an intolerable flicker on the screen. Hence again follows the greater need, for *screen* work, of large coil and current; since the energy required cannot be obtained by slowing and "filling" the spark. All such details must be picked up by personal experience; and gradually the worker will come to depend upon this in his different problems.

For photographic work through difficult subjects, a *mercury interrupter* is thus often of the greatest service. Such an appliance is generally mounted on a separate base-board, but the connections can give no possible difficulty, and the spring is tensioned and slowed down precisely in the same way as the usual appliance. The spark is not only more readily "slowed," but is "fuller" for the same period of vibration, owing to the better contact of the metals. Mechanical contact-breakers are also used by some; but the mercury form, if properly made, answers all purposes, or may itself be worked mechanically at a fixed rate. The chief use of this expedient is to get the utmost photographic effect by switching in the current for one, or very few, extremely powerful sparks only, with a tube of high vacuum. For such an occasional spark a current may be used which would be dangerous to the coil in the ordinary way.

Tubes should obviously, from what has been said, be chosen originally with some reference to the work intended to be done. In particular, the distance apart of the external terminals must be sufficient for the coil. This distance will be ample on any tube, if sparks not exceeding say 4 inches are to be used. But if a 10-inch coil is employed, which may on occasion be pushed up to 11 or 12 inches, and the outer terminals of the tube are only 5 or 6 inches apart, it is obvious that the discharge is liable to spark round the tube from terminal to terminal through the air, especially when aided by any particles of dust on the tube. Large tubes—or rather with good distance between the terminals—should therefore be always procured for large coils, and care taken to keep the outer wall clean, as dust gradually collects by electrical attraction. The platinum wires sealed in must also be thick enough to carry the current, and not be readily cut through by wear at the terminal loops.

With a 2-inch or 3-inch spark, a tube only moderately exhausted will last a long time. With a large coil it is usual to start with at least two tubes, one exhausted to a degree suitable for about a 4-inch spark, and the other for 6-inch or 8-inch. The former can readily be adjusted for shorter sparks by heating; and higher vacua will be gradually attained. Further tubes will be suited to requirements, moderate vacua being as a rule purchased to keep up the stock.

These considerations will clear up questions which seem to puzzle some workers. To take a practical case, a complaint was actually published that an Apps 10-inch coil could only be used up to a 2-inch spark on a certain tube, while a German coil could be used with a 5-inch spark; the result of more was to instantly make the platinum anode red-hot. The stated phenomena really proved two things—(*a*) that the better coil sent as much energy through the tube with a 2-inch spark as the other with 5 inch; and (*b*) that the tube was only exhausted to a vacuum suitable for a " full " spark of say 1½ inch. After a while a longer spark would be borne; but that tube was not exhausted nearly enough to be suitable for starting any ordinary work with a 10-inch coil. Such a case, as Mr. A. C. Swinton has pointed out, also enforces the practical necessity in "building up" a tube (the technical expression for pumping it into proper condition for work) of exciting it during exhaustion with a coil *approximately equal* in power to that with which it is meant to be used. This is a consideration some makers do not seem aware of.

As a rule, with a tube worked to its full power the anode does become of a *dull* red-heat, chiefly in a small spot where the molecules strike, and then by conduction nearly or quite to the edges. Bright red-heat means rather too much current for the vacuum, and to go on with it endangers the tube; but such an extra discharge (*i.e.*, for that tube or vacuum) may be used for a *short* space for some special experiment, or to lower the vacuum as already described. But for continuous work such heating must be avoided, or the tube may speedily be wrecked. The tube itself should be felt from time to time (with the current off), and if very hot the current either stopped for a time or lowered.

92. High-current Tubes.—The original form of focus-tube shown in Fig. 69 cannot be surpassed for work up to 4 or 5 inches of spark, and it is difficult to surpass the *best* specimens with even large currents. Whatever care be taken, however, there is a strange difference in the radiographic power of some tubes apparently similar, difficult to understand; it may possibly have something to do with the radiographic transparency of the glass used. At all events, it is well to know that a super excellent tube, when it has been worked until useless from hyper-exhaustion, generally retains its qualities if re-exhausted. A favourite tube should therefore be sent to be thus renovated when required, as its equal may not so readily be obtained. The tube is more easily exhausted the second time. There is always some difficulty in finding a tube which will utilise to advantage the full power of a large coil. With wide-apart stout terminals, however, and nicely adjusted exhaustion from long use at lower tension, and "well-filled" sparks or the mercury break, some of the most rapid and difficult work has been done with such "lucky" specimens of the plain focus-tube as above alluded to.

Another tube which has been highly spoken of is a German make known as the "penetrator." In this pattern the radiator is not a terminal or anode, and is slightly concave instead of flat: the anode is a ring of aluminium much nearer the kathode, which is of shallower curve than the usual English pattern, and much farther from the radiator. In a separate chamber, is a small piece of palladium, which largely occludes hydrogen, by warming which the vacuum can be lowered — too much and too rapidly in many cases, but this can be avoided by care in manufacture and caution in use. It is an undoubtedly good tube, but I have not found evidence of any distinct superiority over the simpler forms of focus-tubes, of which all modern forms are of course varieties. Some radiographs I have seen appear to imply an inferiority in definition, which I am rather disposed to attribute to the concave radiator.[1]

---

[1] Since the above was in type, I learn that this tube is now made with a plane radiator, which seems as if the manufacturers had reached the same conclusion as stated above. In the last which I have seen, the ring anode is also brought nearer the kathode than formerly, which will of course increase the current to be used with a given vacuum : it may also, perhaps, hasten the period when the tube can be no longer used. Going back to this question of *definition*, my own observations lead me to think that a plane and smooth radiator is of more importance than is usually supposed. Only one photograph has been taken as yet with the tube next described,

If we furnish a tube with a second anode besides the radiator, and the vacuum will only bear a certain current with the radiator alone as anode, more making the latter too hot; by connecting the second also with the positive wire, more current may be passed through the tube. Such a double anode thus gives a ready means of using two strengths of discharge through the same tube and vacuum, which in itself is often an advantage; and it has been assumed that this extra energy produces more powerful Röntgen radiation. It would seem that it must produce results differing in some way; but as regards visible effect upon a fluorescent screen, experimental trials very often failed to show much, or in some cases any gain, with the tubes accessible for trial.

One form which I devised, however, did show such increased effect from increased current, in the most uniform and striking way. It appeared to me that in arranging for powerful discharges sufficient account had not been taken of two well-established phenomena, viz. the need for uninterrupted and smooth *return of the gas-molecules* to the kathode (§ 82), and the greater resistance of a *shorter* distance between the electrodes (§ 84). If we take a series of anodes in a tube, it is the *nearest* which requires the most current to pass a discharge, when the vacuum is at all good; and if the "penetrator" tube had any advantage, it appeared to me to lie in the anode being nearer than the radiator; but no systematic attempt seemed to have

and Mr. Newton thought the definition in it also was not quite equal to good focus-tube work in the detail of the bones. As the kathode rays had in this case passed through two rings (only one, however, being made an anode), before reaching the radiating anode, he thought this might affect it. The fact itself seemed to need more proof: if it be so, the suggested cause is possible, though I think the very uneven radiator a more probable one. It is, however, to be remembered, both as regards the "penetrator" and such double-anode tubes, that in cases where very high penetrative power is required (as for work through the trunk of the human body), any such slight differences in definition become absolutely imperceptible. They are only of importance in a range of work that will probably still continue to be done with plain focus-tubes.

been made to investigate the possibility of improvement in this direction. Messrs. Newton, therefore, made for me a tube containing a series of aluminium rings as anodes at various distances, ranging from about 1 cm. in front of the edge of the concave kathode (which the exceeded in diameter), to some distance *behind* the radiator itself, also furnished with a terminal.

The results were just what I expected, and rather striking. With the radiator alone as anode, the effect was simply that of an ordinary focus tube of similar vacuum, which this arrangement represented. With the nearest ring-anode alone, it was impossible (with this particular and quite moderate vacuum) to pass any discharge at all with a 10-inch coil; rather than pass 1 cm. in the tube the spark passed outside. With the ring alone, representing as nearly as possible the "penetrator"-tube arrangement, more current could be passed than with the radiator, but with no conspicuous gain in fluorescence, or with any other single anode. Commencing with the most distant ring-anode, each was then connected up *along with the radiator* in turn. In every case, as usual, more current could be used without heating the radiator beyond the point adopted as a rough working standard; but no distinct gain in fluorescence was seen (there might be some doubt as to a small percentage—the rough tests did not allow of extreme accuracy) until the radiator was connected up along with the *nearest* anode-ring. With that arrangement, not only could more current be passed—a very large current for the vacuum—without overheating, but this extra current produced a striking, unmistakable increase in the fluorescence.

Two tubes were therefore roughly made as in Fig. 70, adding to the usual arrangements a single ring-anode A, near the kathode (about 1 cm. distant), besides the ordinary radiator. The kathode terminal K must of course be kept wide apart from the nearest anode-terminal C. The same results were observed with these. With radiator alone as anode there were the usual effects of a focus-tube; but when the current and radiation had been brought up as high as the heating of the radiator would allow, by coupling up also the ring-anode c, not only could a great deal of extra current be used, but it resulted in a *proportionate and conspicuous increase in fluorescence.*

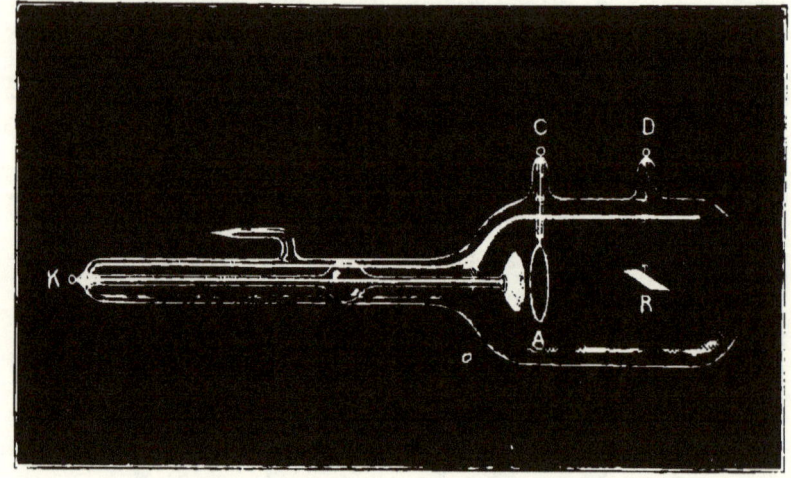

FIG. 70.—Tube with a Second and Close Ring-Anode.

I think it therefore probable that this construction will be found best for utilising the currents of large coils.[1]

[1] The above experiments have been made while this book was in the press, and having been greatly hindered by it, I am unable to give their final results, no really "finished" tubes having yet been made. One point there is a little doubt about: the radiation on the screen from the two best tubes appeared a little more *contracted* (*i.e.* rather smaller "field") than from a focus-tube of usual form. I believe this to be due to the very uneven surfaces of the radiators in each case, but it is possible the ring-anode may have a contracting influence. Neither am I in a position to say as yet that the construction gives more penetration than the same amount of current produces *when it can be passed* into a plain focus-tube; which could however, only be done with a higher vacuum. In any case, however, this particular arrangement of dual anode allows the use of a reasonable vacuum in the usual way; while it will also utilise a powerful discharge when required, with *proportionate increase* of effect. This great range and *elasticity* of effects, with the power of using a powerful discharge before the

One more question is suggested, both by the effects of Becquerel's fluorescent rays, and by the fluorescence of the radiating surface when an ordinary Crookes' tube is used; viz., whether a fluorescent or phosphorescent anode might be superior in radiating power. Professor S. P. Thompson investigated this point, and found that such an anode gave *less* powerful effects, the excitation of fluorescence appearing to absorb useful energy. More recent experiments of Puluj, however, are said to indicate a contrary conclusion. The different results may probably be explained by the fact, already discovered by Becquerel, that *ordinary* fluorescence from light-rays has either no direct proportion or very little to Röntgen radiation; and some experiments have indicated that probably *metallic uranium* might give very powerful radiation. There are several distinct reasons for regarding this metal as very promising. Little has been done, however, with such anodes as yet.[1]

Up to the present, very smooth aluminium is found best for the concave kathode, being a non-volatile metal, and platinum for the radiating plate, whether anode or not. A large concave kathode passes the discharge more readily; but Mr. A. C. Swinton in a paper read before the Royal Society on April 8, 1897, states the remarkable fact that of two concave kathodes on opposite sides of the radiator, at the same distance, but of

vacuum gets very high, is alone a considerable gain; and any slight contraction in the illuminated field, when traced down to its real cause, may I hope prove capable of remedy: it will be desirable to try a ring-anode rather larger and still closer, almost or quite in the plane of the edge of the kathode. · At all events the results appear so promising, that it seems well to put others in possession of them even in their present state.

[1] Puluj has patented a tube with an independent anode, the kathode radiation impinging upon a large flat plate covered with fluorescent salt, this plate being arranged at an angle as usual. The photographic plate or screen is to be arranged parallel to the fluorescent surface. Very powerful radiations have been reported, but such an arrangement could not possibly give the *definition* expected in present radiography. In fact none of the Continental work I have seen has equalled good British specimens.

different diameters, the smaller one (so used in the very same tube and vacuum) causes rays of greater penetrative power. This is obviously connected with the greater resistance of the smaller kathode, and greater P.D. or sparking distance required to overcome it.

93. **Arrangements for Work.**—We may now apply these fundamental facts and principles to actual work ; and in Fig. 71 an entire set of apparatus is represented as set up and arranged for use, with a 10-inch coil.[1] Let us go through it in order. A storage battery is here shown, near the apparatus : if fuming primary cells are used, and placed outside the room, care must be taken to use *thicker* wires or leads than the No. 12 or 14, which suffices here. Between battery and coil are interposed a fusible cut-out, and an ammeter and voltmeter. Neither is absolutely needful merely to do work ; but they are very desirable in order to *certain* work. When long and tiresome experience has discovered the best means (in vacuum, spark, distance, exposure, plate, or screen, &c.) of doing certain work (as diagnosing or photographing the state of some particular organ in the body), the same result cannot again be depended upon, unless *the same amount of current* is reproduced, which is done by reading off on the instruments, and properly adjusting matters, perhaps even adding or withdrawing a cell.[2] If this cannot be exactly done, we can at least *allow* for the known difference in some way—*e.g.*, if at the moment we can only obtain one or two amperes less current than was used before, we may probably compensate this by adding 10 or 20 per cent. to the time of exposure.

This brings us to the coil. It is important to arrange this so as to be well out of the way of all other manipulations, and with the wires connected to the tube as directly and well apart

---

[1] The apparatus was kindly arranged and photographed for me by Messrs. Newton and Co.

[2] A rheostat or variable resistance-coil is very handy, but of course absorbs current, and may need another cell or two.

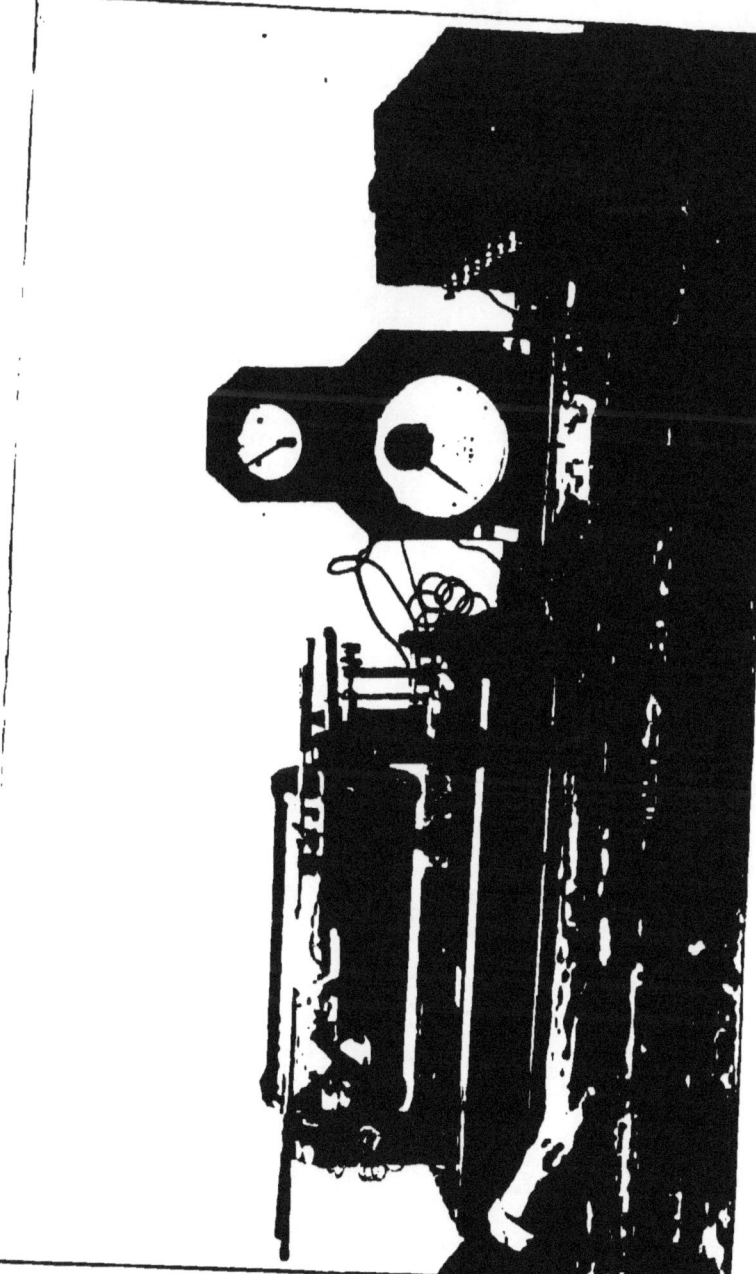

as possible. The commutator or switch should, however, be reached with facility and certainty, for it will be constantly in use ; it is also necessary to reach the discharging handles in order to adjust the spark. The wires from terminals to the tube must be thoroughly and thickly insulated, or unpleasant shocks will be often occurring. In most cases, for these and other obvious reasons, at the back of, or behind the tube, is the best place for the coil. Care must be taken to keep this in good order as respects freedom from dust and damp, or any rusty or dirty contacts.

Get thoroughly into the mind the position of the commutator equivalent to switching *current off* (which in an Apps coil is with the handle pointing upwards), and get the *habit* of regarding that as the normal state of affairs ; always switching "*current off*" except when current is to be passed through the tube, or some adjustment actually tested. If a wire drops from the tube, *switch off* before replacing, or a severe shock will be taken on picking it up : also remember that the coil itself may give shocks if touched or too closely approached near the ends. Ignorant spectators should never be allowed to crowd round the table.

The tube may be supported in any convenient way. A common Bunsen holder often does well, especially for what we may call "small" work, such as photographs of hands or feet. Here we want a square foot or more of firm, flat table surface, on which the plate and hand can be laid, and the tube only needs to be supported over it in such a way, with the anode reflecting downwards, as to leave the greatest possible clear space. For larger operations, such as through the trunk, the best arrangement will suggest itself, and two operators will not adopt exactly the same. A very good plan is to pass the end of the tube a little stiffly through a hole in a piece of cork ; then the cork, with the tube in it, is easily held in anything convenient, more safely than if the bare tube be pinched directly between the jaws of the holder.

Particular care should be taken that the patient be not exposed to shocks from the connecting wires. Any part of the body may, if necessary, be bandaged to a board, to prevent motion during long exposure, both bandage and board being transparent. As a rule it is needless to remove clothing; but if bullets or other small metal objects are being searched for, it is better, in order that buttons or hooks and eyes may not cause any confusion. One more point regarding patients. It ought to be unnecessary to insist that care be taken so as to be unhesitatingly sure about the right and left as represented in a photograph; but cases have actually occurred in which this has not been done, and an incision actually made *on the wrong side!* Such gross carelessness is inexcusable.

Fluorescent screens are for some purposes simply mounted on an upright stand or rod. There are other cases in which a firmer support or resistance is necessary, in order that the portion of the body to be examined may be held immovably in contact. A screen has been laid flat upon a body, the tube being arranged beneath. Such details are merely mechanical, and need not be further considered here. The screen itself is dealt with in § 95.

94. **Study of Each Problem.**—But more is necessary concerning the intelligent adaptation of detail and method (and more especially of tube and current) to the particular problem in hand. It is here where the different character of the rays, and differences in current, exposure, distance, and screens, are of such practical importance. Rays of a certain description are not only better adapted for *penetrating* bodies or tissues of a given character and thickness, but are found specially adapted for *discriminating* between such details amongst different structures. It is also on account of these differences that a coil, giving not less than 6 inches of spark when required, is alone capable of the wide range of problems presented by hospital diagnosis. One or two very simple examples, from various sources, and more especially from the

PLATE II.—FINGERS (*Natural Size*).
Taken with Focus-tube, to show definition of minute Bone Structure.
(*Radiograph by* Mr. T. FRESHWATER.)

published results of Dr. Macintyre, will make at least the leading principles of such adaptation clear.[1]

Consider first such an elementary detail as clear *definition*. A simple subject like the hand, shown in Plate I., when it is merely desired to see the outlines of the bones, with at the outside any injuries, malformations, or perhaps gouty enlargements, may be fairly photographed from an ordinary Crookes' tube, as such were in actual fact first photographed by Professor Röntgen. But the photograph being from the nature of the case not an optical image, but a *shadow* projected on the plate by rays represented by straight lines, one so produced necessarily fails in absolute sharpness, owing to the size of the radiating portion of the tube. The sharpness can only be increased in two ways : (*a*) by increasing the distance of object and plate from the tube, which of course lessens photographic power, but is useful in some cases ; and (*b*) by reducing the radiant to almost a point. This latter method is placed in our power by the "focus" form of tube. The degree of sharpness and consequent detail thus obtainable is shown by Plate II. reproduced of the actual size from a photograph taken by one of these tubes, at a distance of about 9 inches, and which

---

[1] As a matter of simple justice, an English writer is bound to place on record, even in such an unpretending work as this, the leading part taken by an English physicist and a Scottish surgeon in the chief practical advances made since the cardinal discoveries of Professor Röntgen himself; to these latter surprisingly little that is really cardinal has been added since, as already observed. Owing to my long and intimate acquaintance with the leading makers of coils and apparatus, it happened to come within my personal knowledge that at a time when messages from the Continent and New York were daily claiming supposed new advances in tubes and screens, the results of Mr. Jackson's experiments in both had quietly forestalled and surpassed them ; and also when telegrams from Berlin and Vienna were stating as new, photographs or shadows through the bodies of infants, Dr. John Macintyre, of Glasgow, had already sent to various correspondents, without any fuss, similar results through the entire trunk of adult persons, obtained in actual practical diagnosis. The different proportions in actual achievement, and anxiety for publication, are curiously characteristic.

shows distinctly and sharply much detailed structure of the bones.

This is a quite simple case; but, as Dr. Macintyre has pointed out, the problem confronts us, when we deal with subjects of great depth, in a much more complicated way. Suppose we want a radiograph[1] through the entire human body, or through the entire cranium. We have now to consider *what* we want. If we removed the focus tube to a distance proportionate for the much greater thickness, to 9 or 10 inches for the thickness of the hand, our shadow of all the structures would be of approximately equal sharpness. But in the first place, such a distance will afford far too feeble radiation; and in the second place, even supposing that we could thus get shadows of all the structures super-imposed, these would only confuse each other. What we want is some definite portion, the less confused by images of the rest, the better. These objects are best attained by placing the focus tube as close as possible to the structures we *do not want*. If we seek for a suspected injury or disease of the bone on one side of the cranium, therefore, we bring the tube close to the other side, and the photographic plate close to the side we wish to photograph. Then the shadows of the side next the tube are so diffused and dispersed and enlarged that they practically disappear; while the details are sharp of the side in contact with the plate.[2]

In the same way, taking a photograph through the trunk, if the subject lies on his back upon the sensitive plate, and we bring the focus tube almost in contact with the chest, the spine will be photographed, while the sternum and front ribs disappear. Reversing the position, the spine is practically got out

---

[1] This seems, upon the whole, the least objectionable of the various names which have been used for photographs by Röntgen rays.

[2] The possible physiological results must, however, be borne in mind, of too close a tube (§ 100). Temporary loss of hair has been reported from exposure to too powerful radiation.

of the way (except so far as its opacity or *resistance* is concerned —see further on) while the sternum and ribs are sharp. The same position is required for the heart, or other non-osseous intermediate structures; but in this case we must further seek aid from the character of the rays. In regard to *definition*, however, the general rule is obvious; to get the organ or structure desired, as near to the plate and as far from the tube as we are able to manage. And as our distance is limited by the power of the radiation, and how far it will bear dilution, this consideration tells powerfully in favour of large coils.

But we have further to adapt our radiation to the *nature* of the structure. It is here, perhaps, that on the one hand the greatest advances have even already been made; and on the other, that we have still to expect most from further investigation. Only main general principles can be said to be as yet ascertained; and it is possible that even methods of photography may be found to give valuable differences in contrasts of shadow for different structures, as well as modifications in current and screen.

The general law appears to be, that taking a simple subject of small thickness, such as the hand of Plate I., we must use a comparatively small current and low vacuum to obtain opacity of the soft tissues; while a longer spark and somewhat higher vacuum gives transparency of these, and comparative opacity of the bones; and if we push spark and vacuum still higher, even the bones give very little shadow at all. But, if the whole trunk is the object concerned, the question is more complicated : for the great resistance to the rays *compels* us now to use high vacuum and powerful spark to overcome it. It has, however, been found that, even in this case, modifications in the vacuum and spark affect the photographic or fluorescent result to an extent which may be very useful; one vacuum bringing bones into more prominence, while another state of the tube will bring the softer tissues into more contrast of detail.

The *general* character of the means to be employed for a

good result can be pretty well known beforehand. Thus, such a subject as Plate III., reproduced from a radiograph of a small snake[1] taken by Mr. T. Freshwater, could not be dealt with effectually, at any ordinary distance, by a powerful discharge in a tube very highly exhausted. There would hardly be any shadow whatever perceptible. It was taken as a matter of fact with a spark somewhat under 1 inch, and an exposure of 10 seconds. Whenever such a subject has to be attacked with a large coil, either the spark must be reduced, or the distance of the tube greatly increased, or (in case of photography) the exposure reduced to a minimum. It is quite obvious that in either case the energy of the rays is lessened; but the interesting (and most ultimately hopeful) fact is, that the effects of reducing the energy in these different ways are *not all alike*. While the bones of the snake will be best shown by a diminution of spark and vacuum, short exposure, and possibly increase of distance if also necessary; the bringing out of softer structures is more likely to be effected by combining as small spark as possible with longer exposure. Under-exposure may often be "intensified" by the usual methods; and such under-exposure so intensified sometimes gives different and better results still; as may also long and patient development.

Taking, however, a subject like Plate IV. (*Frontispiece*), from a radiograph obtained by Dr. MacIntyre [2] in a case of exten-

---

[1] The animal was of course chloroformed.

[2] I selected this radiograph as one of considerable historical interest, being the very first really successful one of the difficult pelvic region, and taken so early as the spring of 1896. No printing process can quite reproduce all that may sometimes be seen in an original radiograph; but the following, which I am allowed to quote from a work still in the press when this is written (Dr. Owen's *Surgical Diseases of Children*) may be interesting as showing what professional eyes read even in the print here given. "On the sound side (to the spectator's right) the line of the temporary cartilage is shown in the floor of the acetabulum, and the rim of permanent cartilage, the cotyloid ligament, is also seen. There is, moreover, a faint trace of the layer of cartilage joining the head of the femur with the diaphysis. The outlines of the neck and of the great trochanter are very definite. On the

PLATE III.—SMALL SNAKE

Illustrating subject for sh

sive "hip disease" through the entire pelvic region of a boy, we *must* use for such, a powerful discharge in high vacuum, in order to get through the enormous shadow-resistance. This was taken with a 6-inch control discharge from a 10-inch coil, with an exposure of fifteen minutes, which was, however, intermittent, and probably not exceeding ten minutes of continuous discharge. Less than a 6-inch coil would be no use, and it is in such cases that still larger coils are especially useful. It remains to be ascertained, however, whether or not for some such cases a much greater *quantity* of current at a lower *tension*, especially if combined with a large cup-shaped kathode, might give better results; such discharges as would be obtained from a coil wound with much thicker wire, but supplied with much more primary current from a dynamo. It may very possibly prove that such variations from the hitherto usual proportions of Induction Coils might give more efficiency in some problems of radiography, just as a 2-inch spark from a 2-inch coil, and a spark lowered down to 2 inches from a 10-inch coil, are very different in their effects.[1] It is at least probable that two different primaries, or one capable of being used in two ways, as in the great Spottiswoode coil (p. 37), would prove very useful for large coils in radiographic work.

However this question may ultimately be determined, the

affected side the Y-shaped cartilage and the cotyloid ligament have been entirely destroyed by the growth of granulation tissue, which has also caused absorption of the head and most of the neck of the femur. The result of this femoral absorption is that the great trochanter is inconspicuous, and that it is considerably raised above the normal level. The destruction of the Y-shaped cartilage has determined a premature consolidation of the os innominatum, which is seen to be considerably smaller and shorter than its fellow. The long-standing articular disease has moreover had a prejudicial, trophic influence upon the upper part of the femur, which, just below the trochanter, appears wanting in solidity and strength. There is, in fact, an obvious dwarfing of femur as well as haunch-bone, and the measurements of the affected limb would show a very considerable and inevitable shortening."

[1] Dr. Macintyre has used as much as 30 amperes of primary current.

greatest difficulty is to get details of the *softer* tissues, through a great *thickness* of the body; where the amount of obstruction to all rays makes it impossible to use the lower current and vacuum which appear in themselves most suitable for such softer tissues. The one thing clear is that the part wanted must be placed as far from the tube and near the plate as possible, both for sharpness, and that the energy may be generally reduced *before* reaching the part desired. Beyond that every operator must work out the actual details for himself; because the exact *data* of his own particular coil and primary current will affect all the other *data*, in manner not yet altogether understood. Where long exposure can be given, it will make the task easier; thus, one operator reports excellent detail of the muscles of a rabbit, and tendons of a partridge, with a "distance" of 14 inches, spark a little under 3 inches, and an exposure of 75 minutes. But with the trunk of a living human subject, such an exposure could seldom be tolerated. On the other hand, this class of work is particularly suitable for *intensification*, or for long and gentle development, which seems to work especially well in bringing up detail of the softer tissues. It is always advisable, in dealing with structures which, from their position in the body, cannot be brought close to the photographer's plate, to work with the tube at as great a *distance* as the means at disposal will allow, in order to secure definition; but unfortunately this expedient is least available just in those massive subjects for which it is most wanted. So we are again thrown back upon whatever may ultimately be found the most powerful and suitable kind of discharges.

Where the subject differs considerably in thickness, a little manipulation may be needed to equalise matters. Suppose, for instance, that hand and arm are desired in one photograph; the hand is much thinner than the arm. In such a case a better result may be obtained, as Professor Morton suggests, by waving a leaden screen to and fro between the tube and the hand, so as to intercept some of the radiation through the thinner portion.

There is yet another means of effecting a sort of *self-adjustment* for the special problem, to at least a certain extent. After penetrating a subject, the energy of the rays is so much weakened, that a glass photographic plate absorbs it materially. Therefore, by placing two or three plates above each other, either in contact, or with sheets of black paper or foil between them, a similar number of photographs are obtained, with very little difference in sharpness, but due to gradually lessened amounts of energy. There is very great probability that some one of these results will be perceptibly better than the rest, as regards the special object in view; and this may give valuable hints for future proceedings.

95. **Fluorescent Screens.**—In hospital work and diagnosis a good screen is as important as a good tube. Here again Mr. Herbert Jackson was the first to produce one of real efficiency, and which is still extensively used, though others are often employed. This consisted of potassium platino-cyanide, a coat of small crystals being made adherent to a paper screen; it fluoresces very brightly of a pale blue. The salt crystallises with three proportions of water, the most aqueous giving the best result; hence loss of moisture is apt to produce some uncertainty as to the amount of action, which is best counteracted by keeping the screen in a damp place. The platino-cyanide of lithium, and the double salt of potassium and sodium, have also been used.

Professor Röntgen's first experiments were made with platino-cyanide of barium. This will show the bones of the hand very readily, but in general is inferior in mere brilliancy to the potassium salt. But the curious fact is observable that different salts are as *discriminating* in their results, as differences in current and tube; so that even ordinary samples of the barium salt, though less bright, sometimes show more *contrast* in shadow effect than the potassium. Much also depends upon the degree of purity, and the exact molecular condition. Powder is with most substances far less effective than natural crystals, and by purifying and repeatedly re-crystallising the barium salt, finally producing crystals small enough to resemble meal or powder, the brilliance is greatly increased, and the definition improved. Where *definition* is especially desired, the barium salt thus prepared is preferred by most to anything else, and has the advantage of permanently standing a great deal of use. The best specimens are now quite equal in brilliance to potassium screens.

Other compounds will, however, do very efficient work, and have been employed by various operators. In America calcium tungstate is largely used, following Edison, who, after trying many substances, selected and preferred this compound, working upon the natural mineral product. In England it is not thought equal to the above; but again, the quality of the sample has very much to do with the effect. One operator, who speaks highly of it, prepares it by precipitating sodium tungstate with calcium chloride, washing perfectly and drying at a gentle heat, then fusing (in a cavity made in charcoal) under the O. H. blowpipe, making it boil for a few seconds, and finally cooling it very gradually; the mass is powdered and applied on wet gummed card or paper to make a screen. What chemists term a "solid solution" of copper tungstate in calcium tungstate is also said to be very effective. Iodide of rubidium has been reported as considerably surpassing calcium tungstate, but the scarcity of this metal has prevented much experience of it. The double fluoride of uranium and ammonium also makes an excellent screen, which some think equal for average purposes to any other.

Other efficient materials are likely to be discovered; but it is not likely that any one will be able to be regarded as the "best" screen for all purposes alike, for reasons already indicated. Most of the best known radiographers employ at least two screens. There appears also much to be learnt yet as to the best way of preparing a screen. Dr. Macintyre discovered that no screen yet made would absorb nearly all the energy: after producing an image on one screen, there were sufficient rays which passed through to produce another image on a second, 12 inches behind it; and yet another upon a third, at similar distance. Hence he prepared a screen with much coarser crystals, giving a correspondingly thicker layer of fluorescent salt, with very appreciable gain in the brightness of the image. But unfortunately such coarse crystals give an image of less definition.

The fluorescent material is generally spread upon a sheet of paper—preferably black or blue paper—stretched in a frame, and placed with the paper towards the tube, and the fluorescent material towards the eye. The screen may be mounted so as to form the further end of a box shaped as the frustrum of a pyramid, with eye-lenses or simple eye-holes in the smaller and nearer end. Such an arrangement is in America termed a *fluoroscope*, and is in some respects very convenient, but many workers prefer the plain screen and a darkened room.

Screen work stands by itself in one respect. By it alone can the *motion* of organs be seen. With a good coil and screen there is no difficulty at all in seeing the beating of the heart, movement of the diaphragm, and some other phenomena in the human body. The clothes need not be removed; and it is curious to see the black shadows of buttons on the garments, even the small thread-holes in their centre being often observable.

96. **Photographs of Screen Images.**—As early as March, 1896, Battelli and Garbasso took photographs with a camera of the image on a fluorescent screen. This is of course possible; because the unrefractible Röntgen rays are there converted into visible fluorescent rays. Practical difficulties caused this method to be lost sight of, but Dr. Macintyre has revived it with some success. The chief difficulty is that, supposing

we place the camera and lens some feet behind the screen, while the lens gives an *optical* image of the screen only, the Röntgen rays which (as we have already seen) get through the screen, give another *shadow* image of the lens-mount and other metal on the camera-front. This Dr. Macintyre got over by covering the camera-front with a sheet of lead pierced in the centre by a hole for the lens only. A good image is thus obtained; and the advantage of this process is that it adapts itself to the most rapid plates and processes, and enables large subjects to be reduced directly to the size of lantern-slides for public demonstration. Another considerable advantage is that, as already indicated, it is generally much more convenient to direct the rays through the subject in a horizontal direction, the management of the vacuum by heat under the tube being then easily effected. When used in this way there is more photographic effect from the potassium salt than from the barium; but the definition is not so good, and the best barium screens are bright enough to yield very good results.

97. **Stereoscopic Photographs.**—Effective stereoscopic double photographs were first taken by Professor Elihu Thomson, in the United States. The method is very simple. The hand or other portion of the body is laid upon a wooden frame, underneath which a wooden slide enclosing a plate can be introduced and withdrawn. The tube is mounted so that it can be moved horizontally 2 or 3 inches and clamped. One plate is introduced, a photograph taken, and the plate withdrawn. The hand being kept in the same position, the tube is shifted, clamped in the new position, another plate slid underneath, and the other picture taken. Some of the results are rather curious; and possibly, when greater advances have been made in photographing the softer tissues, this method may assume more importance. At present its most interesting applications have been made by Messrs. Remy and Contremoulins, who, by injecting the veins and arteries of some severed member with a solution of sealing-wax, in which bronze-powder

is diffused in suspension, have obtained stereoscopic views of the system, in all its space relations, such as are possible in no other way.

98. **Kinematograph Movements.**—The last development of radiography, up to the present, has been in the direction of recording the movements of parts of the body or organs within it, by successive pictures exhibited by a kinematograph. As most readers will be aware, the latter is an instrument for rapidly placing in succession before the eye pictures of moving objects taken at very short intervals—at least several, and usually many, in one second—arranged in a long strip, the effect of which is that the successive phases of the motion are apparently reproduced. Dr. Macintyre has been recently endeavouring to secure this result, and obtained results both promising and interesting, some of which were exhibited at a meeting of the Glasgow Philosophical Society in March 1897, when all the movements of the knee-joint of a frog were well demonstrated to the audience by means of an ordinary kinematograph.

Success necessarily depends upon rapidity of exposure. For details of the precise means employed, reference must be made to the 1897 *Transactions* of the above-named Society.[1] Early in 1896 Dr. Macintyre found that with a large current and the full spark of the mercury interrupter, one flash of the tube would give a photograph of the bones of the hand; and with ten flashes an excellent picture was obtained, showing the minute structures of the bones. In these ten flashes, of course, the swing of the vibrator occupied much more time than the ten actual discharges; and the problem chiefly is to reduce such idle time. This was accomplished to a considerable extent by using a rotating mercury interrupter driven by a small electric motor; a wheel rotating at great speed causing the fork of the interrupter to dip with great rapidity into the platinum and mercury amalgam (*see* §§ 19, 91). He antici-

[1] The paper is not published at the date this is written.

pates still better results with a purely mechanical interrupter, also driven by a motor.

Two general methods were tried. In the first the shadow was formed on a fluorescent screen, and by the camera ordinary photographs were taken, reducing the pictures to the size required for the strip; but this plan up to the present has proved too slow. He next covered the kinematograph with a plate of lead, in which was a small aperture covered with black paper, corresponding to the aperture behind which the strip of film passed. The frog was selected for this experiment, the animal's leg being placed over the aperture. In the present experimental stage, the best results were obtained by fixing the limb so that the movements could be controlled by the operator, and still further controlled by a little anæsthesia.

It will be quite obvious that the small area for visible movement in this latter method, and the present lack of sufficient energy for the camera method, present great practical difficulties in this branch of work. But success would open out so many valuable possibilities, that it is to be hoped they may not be found insuperable.

99. **Photographic Exposures.**—It is beyond the purpose of this hand-book to enter into details of the photographic work, concerning which there are many admirable manuals. It will suffice to say that the experienced photographer will do well to begin, at least, with the plates and processes he is familiar with, bearing in mind what has already been said concerning the occasional great delineating value of *under-exposure with intensification*, or with long and slow development. A beginner will perhaps do as well with Cadett's "lightning" brand as any, using generally the Velox developer. I have also seen splendid work with Paget plates; and have in fact no reason to think that any standard brand will fail to give good results when properly managed. It has been stated that fast and slow plates give similar results with Röntgen rays. That is certainly not the fact stated so broadly; but it is true

that the time-differences are not, apparently, nearly so great as in ordinary photographic work, and some medium plates, at least, will do even quicker work than the rapid ones of the same make. It is probable that here also greater adaptability of certain plates to certain structures may ultimately be found.

M. Vandevyver considers, after many experiments, that the necessary exposures for bodies or limbs of various thicknesses, vary about as the *cube* of the thickness, other conditions being equal. I doubt this, as also another conclusion he announces, that the radiating energy from a tube diminishes only as the distance, and not as its square. In any case it has to be remembered that the conditions are generally very far from equal, a much more powerful discharge through a higher vacuum being necessary when great thickness is concerned. Also that long exposures with (comparatively) small spark give quite different *qualitative* results, from a large spark with shorter exposure.

The most interesting photographic question is, perhaps, how far it is possible to shorten the exposure (a most desirable object with living subjects) by using a fluorescent screen in contact with the sensitive film. Considerable effect of this kind has been demonstrated by many different workers; but the great difficulty has been a certain amount of diffusion in the image, caused probably by a spreading of the effect through the crystals, which have generally been used on the fluorescent screen. This appears to be best avoided by causing the rays to pass *first* through the glass plate, then to traverse the film, afterwards striking on the fluorescent surface, in actual contact with the film.[1] Winkelmann and Straubel have lately reported that with this arrangement, and using fluor-spar as the fluorescent material, the photographic film is affected almost a hundred times as rapidly as when the plate alone is employed.

[1] The glass plate itself, in this case, obstructs the energy considerably. This is easily avoided by using instead a celluloid film, which is radicgraphically transparent.

As ordinary fluorescence seems to bear no very direct relation to the amount either of Röntgen fluorescence, or Röntgen radiation, so also it seems to stand in no invariable relation to the amount of photographic acceleration; and there would, therefore, appear to be a wide field for experiment in this direction.

100. **Physiological Effects of the Rays.**—There is ample evidence that prolonged exposure to powerful Röntgen rays may have very serious effects upon the living body; different individuals, however, being variously susceptible to such effects. Most persons are conscious of peculiar feelings in the hand whilst the latter is being photographed with a powerful coil; but this is a transient feeling. It is when the hands or other portions of the body are in the path of the rays for long and frequent periods, that serious results are likely to follow. Dr. Macintyre reports severe dermatitis (inflammation of the skin) as the result in his case, the hands becoming red and swollen, followed by shedding of the epidermis and loss of hair. But a demonstrator who was engaged several hours every day for some months, experienced far more grave consequences. Here again no serious inconvenience was felt for several weeks; but gradually small dark blisters appeared under the skin, which became very irritable, with much inflammation and pain. This was allayed by ointment; but the skin became hard and yellow, like parchment, and peeled off; the same symptoms being repeated with the new skin. Then the tips of the fingers swelled, feeling as if they might burst; the nails became affected, from under them came a copious discharge, and the nails came off. The hand then had to be kept in bandages for weeks. Inflammation of deeper structures has also been reported. Recourse was had to anointing the hands with lanolin and wearing kid gloves over, which also became saturated with grease (the gloves were of course transparent to the rays themselves), and it was thought that considerable alleviation was thus obtained.

Protracted exposure of the hands during manipulation should therefore be avoided as much as possible ; ordinary, occasional, or brief exposures will give no trouble, unless perhaps in some abnormally sensitive subject. Should protracted exposures be necessary, protection might probably be obtained by kid gloves covered on the back with lead foil, which could be kept in place by another thickness of leather stitched over.

This powerful action of the rays suggests that they may possibly be found useful in the treatment of various diseases, and more especially of the skin. Reports from America have stated that they kill bacteria ; guinea-pigs being inoculated with both diphtheria and tuberculosis, and those subjected to the rays remaining free from disease, while the others all succumbed. This cannot be said to have been authenticated in any satisfactory manner ; but the really fundamental fact of most energetic physiological action, from a form of radiant energy which is capable of penetrating every tissue of the body, is certainly fruitful in encouragement and suggestion.

101. **Starting a Coil.**—As this little work may come into the hands of some who are using an Induction Coil for the first time in this kind of work, it may be useful here to explain, simply and consecutively from the first, how to get it into work. For the battery power suitable for various coils, reference may be made to Chapter III., and for this especial class of work we will only repeat the advice to use secondary batteries if possible. Read also again carefully, § 26 in that chapter concerning " personal precautions." We will suppose a 10-inch Apps coil, and half a dozen lithanode cells, which will give at starting about 12 volts and nearly 10 amperes, a current suitable for such a coil. The battery cells will be marked P and N, or + and −, and the battery terminals of the coil are marked P and N. With a primary battery, the platinum or carbon plate will be the P pole.

Now to proceed. *Switch off the commutator* (this is italicised throughout, because from the first the *habit* of it should be formed), which in an Apps coil is to point the lever handle vertically upwards. Using thick wire, as directed in Chapter III., connect P or + end of the battery with P on the coil, and − or N with N, with a fusible cut-out if storage cells are used; and the measuring instruments in the circuit if these are used. Here note that on an Apps coil the marked tops of these terminals, PN, are movable and interchangeable, so that either terminal may be made P. This is in order to arrange the current wires so as to connect without crossing, wherever the battery may be placed. Consider therefore the best arrangement of battery, and then arrange the terminals accordingly, and keep them so. It does not matter in the least which way they are arranged; but what does matter is that they should be kept the same, so that the commutator may be switched on *one way* or the other with certainty.

Next set the dischargers. Reckoning them as facing the operator, draw the right-hand one as far back as possible, and turn it round at right angles, or with the handle towards you and slanting upward. The left-hand one push in as far as it will go, pointing to the knob through which the right-hand one slides. This secures a discharge between the left-hand *point* and the right-hand *knob*, which is the proper condition of sparking terminals. (For subsequent discharges in work, a knob or disk is better placed on one of the points, which should then face each other; but with small sparks it matters comparatively little.)

Next see to the interrupter. The coil being supplied by any one who understands the matter, will be found to have all the tension taken off. If not, turn to the right the screw T (Fig. 16, p. 30) till the iron hammer bearing one platinum contact is as close to the core of the coil as it will go; and screw up the top milled-head B till the other platinum contact barely touches the first, leaving scarcely the thickness of a sheet of paper between the contacts.

Now turn down the lever of the commutator to a horizontal

position on one side or the other, which switches on the current. The condenser and secondary will be already properly connected by the makers. The current being now on, turn the screw T of the interrupter *to the left* gradually, which tensions the spring and draws the hammer back from the core. The hammer will soon begin to vibrate, and small rapid sparks pass between the contacts, but nothing probably passes the discharger. Turn the screw farther to the left, and at length the secondary spark will pass. As soon as it does, *reverse* the handle of the commutator, through the *off*, to the opposite *on* position. In one of the two positions it will be found that the secondary spark passes much more freely. *That is the position* in which henceforth to switch on the current, with the present arrangement of the battery and discharger terminals. This is the readiest, and a nearly instantaneous method of finding the proper direction; in which also, with these short sparks, a faint "brush" discharge will be seen to proceed from the left-hand point, towards the knob on the right.

We may now look at a tube, supposed to be an unknown tube, but believed to be suited for a moderate discharge from our coil. First *switch off current;* and it is better now to turn th two discharger-points towards each other, but putting a knob or disk on the negative one. The tube being mounted in its holder, connect the right-hand discharge-terminal (the knob in the foregoing operations) with the *cup-shaped* aluminium kathode of the tube, the other with the anode; using of course finer wire than from the battery, but still perceptibly thicker than that of the secondary coil. These connecting wires should be well covered, and in short enough lengths to avoid useless loops or coils, which might occasion unpleasant or dangerous shocks. All being connected, switch on the current. The discharge will probably not pass in the tube, but through the discharger as before. *Switch off current*, and withdraw the left-hand discharger to a further distance—say an inch more with this coil, or $\frac{1}{2}$ inch with a 4-inch coil. Unscrew the platinum contact a little further back by screwing B (Fig. 16) to the left, not however increasing the distance between the contacts beyond about $\frac{1}{32}$ of an inch; switch on the current again, and turn the tension-screw T gradually more to the left, till either the tube fluoresces or the spark again passes dischargers. If the latter, *switch off current;* withdraw the left-hand discharger to its greatest distance (which will be over the "guaranteed" spark); switch on current again, and further turn the tension-screw of interrupter to the left, till the discharge passes in the tube, and the latter fluoresces a bright yellow-green from the "kathode" rays.

Every time anything is done, except to tension the interrupter, *switch off current* first. Even the handles of the discharger should not be manipulated with the current on when any but a very moderate spark is arranged, or a powerful shock may be felt, the insulation of the handles not being enough to stop a powerful discharge, which may also spark back to the hand from the metal parts of the discharger.

Notice also the ammeter, while the interrupter is being tensioned; as the tension increases, slowing the spark and allowing time for more primary current to pass the contacts, the amperage will be seen rising. The use of the voltmeter is to show the gradual decrease of battery current. A slight fall may be compensated by either a little more exposure, or with storage batteries, by tensioning up for more amperage. When the full voltage of storage cells falls to about 80 per cent., the cells should be exchanged or re-charged.

We have now simply brought the spark up to discharge through the focus tube; the discharge has still to be *adjusted*, according to circumstances. It may be already obviously too powerful for the work wanted; in that case we try another tube, or *heat* this one as already described, lowering the tension and shortening the spark to correspond. By the interrupter alone, a large coil can thus be brought down to a discharge of even a quarter of an inch or under, with the full battery in

STARTING A COIL

circuit. At the same time, this is not good practice; and when low-power work is really foreseen and intended, it is far better to bring only a few battery cells into the circuit.

More probably, however, we may want to increase the power of the discharge, which can be done to a great degree beyond what will first pass the tube, by further tensioning the spring of the interrupter. It is well first to ascertain what the discharge is; to which end *switch off current* and shorten distance of dischargers; repeating successive adjustments till the distance is found at which the discharge *just* passes the tube in preference to the air-gap. From this one can generally approximately judge a sparking distance that will give about what is wanted, which is then worked up to as before. Or we may leave the dischargers wide apart, and simply work up the spark to as much as the tube will bear, or which gives the best and nearest result to what we want on the screen. Then the discharger-distance, which measures this, can be ascertained. It is therefore time now to examine the effect on a fluorescent screen, arranged between tube and observer, with the prepared side towards the latter, about 9 to 12 inches from the tube, the latter of course so arranged that the anode reflects rays to the screen, and holding the hand or other object close to the screen on the other side, in a dark room (unless mounted as a fluoroscope). The effect can thus be tried of still further tensioning the spring. If the effect desired is *better* seen, the moral is obvious.

The increased energy will now perhaps make the anode quite red-hot. This may be borne for short periods; but a tube should not be worked long at a time with a bright red-hot anode, or it may break down. If such powerful discharge is needed, it is better to have recourse to a higher vacuum, which will stand it without such heating of the anode, or to a tube with a second anode as described in § 92.

For the utmost work the coil is capable of, it may be worked for a few minutes with the dischargers at their furthest (which will be $11\frac{1}{2}$ to 12 inches with a 10-inch coil), and a tube must be used which has become too highly exhausted for ordinary tasks. Then, not only will the interrupter be tensioned up, but the platinum contact must be still further screwed back at B (Fig. 16), till the sparking distance may be $\frac{1}{16}$ of an inch. or even more. This will slow and increase the "quantity" of the sparks, forcing a discharge and giving effects which can be obtained in no other way. Such long sparks, however, rapidly injure the platinum contact surfaces, and should only be used on really necessary occasions; in fact, the mercury interrupter, or a mechanical one, is much better for this class of work.

The contacts of the interrupter require careful smoothing from time to time, with a file supplied for the special purpose; at intervals they need renewal. They last longer if the primary current is reversed at moderate intervals, so using them equally each way. There are two methods of doing this; in each case it must not be forgotten at the same time to reverse the wires connecting up the terminals of the tube. (This is another reason for always fixing a knob or disk on the point of the negative discharging-rod, in order to *mark* the terminal in use as the negative one.) One method is simply (having reversed the terminals of tube) to *reverse the commutator;* this is simple, but has the objection that it prevents forming an invariable habit of always switching on the commutator *one way*, which in many respects is desirable. On that account it is better perhaps to reverse the *wires from the battery*. Then an invariable method of handling the commutator is not interfered with.

When work is over the interrupter-spring T (Fig. 16) should always be slackened, and not left on tension. The satisfactory working of the coil depends greatly on this spring preserving its qualities, and to do this it must be relaxed when work is over. Tube, and sparking distance of dischargers, however, when once adjusted, are often best left as they are, ready for the next experiment. Then the spark will only need to be tensioned up again to the required point.

Should at any time the sparking distance of the coil greatly fall off, while the spark at the interrupter greatly increases, it is probable the *condenser* has been pierced by a spark. On this see p. 56. A coil often worked should always have a spare condenser at hand.

102. **Easy Experiments.**—It may be well to state, for owners of small coils, that a number of easy subjects are well within the power of an inch spark and a small focus tube of suitable vacuum (which may probably have to be warmed down to the proper point). The snake (Plate III.) is such a subject, or a small flat fish, or even a child's hand, or a small frog. Purses, containing coins and other metal objects, or such objects placed between two thin boards, are easily photographed by such a coil. Hence experimental demonstrations, sufficient for mere intelligent interest in the subject generally, may be indulged in at very small expense.

103. **Vegetable Physiology.**—It is also well to mention that while this chapter, for obvious reasons, has given sole prominence to the use of Röntgen rays in physiological, pathological, and surgical work, they produce equally interesting results when brought to bear upon the tissues of plants. These results may perhaps prove to be even more generally interesting to non-professional students. In some respects this class of objects are easier to photograph, and in some more difficult; there being as a rule not such a combination of very opaque with very transparent portions, as are often superposed in, say, the human thorax. Such subjects, again, are much more easily manipulated and more variously prepared; and many of them lend themselves better to moderate discharges and long exposures.

104. **Dynamo Currents.**—Only a few words are needed here respecting the use of Induction Coils in connection with these, because in every case the arrangements will have to be

made by a competent electrician, who will adjust the current to what the coil will bear. A rheostat or variable resistance-coil, or a water-resistance, which is handy and cheap, will be absolutely necessary in the primary circuit, as well as the measuring instruments. Where such a current supply is available, it is much cheaper, and saves a great deal of trouble with batteries.

A *continuous* current will be sent through the coil just as usual, only requiring to be adjusted as regards its amperage and voltage. This kind of current is much the best, and the employment of such large dynamo currents is of considerable interest, in connection with a question already mentioned, as to whether or not heavier primary currents with thicker wires in the coils may place additional resources within reach of the radiographer. Future editions may be able to record the determination of this point.

An *alternating* current, which is most usual in public supplies, has one advantage in being much more readily modified by a supply "transformer" to any dimensions desired. On the other hand, all the modified effects (§ 18) of the "condenser" are lost. The electrician having adjusted the current, it will be connected direct (except for the rheostat and measuring instruments in the circuit) in the case of an Apps coil, with the terminals marked "ALT" or "COIL," simply disconnecting the condenser by withdrawing the wires. In other coils, the condenser must be disconnected, the tension-spring slackened, and the contacts screwed close up. That is all which is necessary in regard to the coil.

But the tube and its management must be modified, as we have to confront two opposite discharges of equal strength. Platinum is a volatile metal when it forms the kathode in a high vacuum; therefore an ordinary focus tube will not answer. Mr. A. C. Swinton has used an elongated bulb with two opposite concave terminals of aluminium, both supposed to be "focused" upon the same plate (not an anode) of platinum.

This is quite correct in theory, and utilises both the opposite discharges for radiation. But both have to be sealed into the glass, and in practice it is found impossible really to adjust them so as to focus both on the same point; hence this form fails to give the sharp definition now demanded in good radiography.

It is better on the whole to sacrifice half the energy by using such a tube as shown in Fig. 72. Here A and B are the two

FIG. 72.—Tube for Alternate Current.

terminals, with cups of aluminium, each of which is alternately kathode and anode. The kathode rays of each impinge upon the platinum radiating plate R, which reflects one discharge out on one side of the tube, and the other on the other, only one being used. We lose half the energy, but we insure radiation from one point only, as in the usual form of focus-tube.

The radiator R may be connected with another terminal C. When thus provided, the same tube may be used with R as an anode, in the ordinary way, whenever continuous current is used with the coil; or if a branched positive wire be connected with both c and either of the other terminals, we then have a tube for such use with a double anode, as described in § 92, and with more or less of the advantages there mentioned.

Tesla Transformers behave essentially as Induction Coils used without condensers and with alternate currents. Their variety of proportions invite experiment in regard to the effect of different currents; but details do not belong to the subject of this handbook. At one time they were occasionally used in the *secondary* circuit of an Induction Coil, as a supposed improvement. That idea is now quite abandoned, it being well established that the best result is obtained from an ordinary coil when it is used alone.

₊ As these final pages go to press particulars reach me, in a paper by Messrs. Sayer and Willyoung in the *Journal of the Franklin Institute* (New York), for March, 1897, of an improved contact-breaker, by which it is claimed that heating is greatly reduced, and sparking at the contacts almost abolished, leading to great increase of power for the same current and coil. Briefly, this is accomplished by mounting the swinging hammer apart from the also swinging contact-piece, so that it swings a certain distance *before* carrying with it the contact-piece. Thus the latter, not being acted upon at what may be called a dead-point, but caught away by the swing of a mass in motion, is removed from contact far more rapidly, causing less sparking and more inductive effect. There is of course some modification in the " make " as well as " break," which cannot be discussed here. It is remarkable that almost simultaneously similar results have been claimed in England for a contact-breaker devised by Mr. James King, known as the " Vril," " Acme," and perhaps by other names. This contact-breaker, however, employs somewhat different means.

END

BY THE SAME AUTHOR.

Second Edition. Crown 8vo, 7s. 6d.

LIGHT: a Course of Experimental Optics, chiefly with the Lantern. By LEWIS WRIGHT. With Illustrations. Second Edition. Revised and Enlarged. Crown 8vo, 7s. 6d.

NATURE.—"A book by a worker whose work in his own line is of a very high order, and whose experience will be of correspondingly high value to others who are working at the same subject. . . . His book is a valuable repertory of useful information and of suggestive hints. . . . The numerous illustrations, a large proportion of which are original, add greatly to its value. The coloured plates of polariscopic phenomena are, it should be added, of singular excellence."

MORNING POST.—"Mr. Wright's book has one special and decided advantage over the majority of works written with the object of instilling scientific knowledge into the minds of readers who have not had a scientific education. It is easy to understand. . . . Another recommendation of the volume lies in the fact that all, or nearly all, the experiments described can be reproduced by the reader, without any extravagant outlay in the way of apparatus."

GUARDIAN.—"A very good book for the practical study of the subject."

JOURNAL OF THE ROYAL MICROSCOPICAL SOCIETY.—"A book on which the author may be very much congratulated, as in our view it is by far the most useful work on its subject to which the general body of microscopists can refer."

DAILY NEWS.—"A popular scientific treatise deserving of commendation by reason both of its strictly experimental character and of the simplicity of the modes of practical demonstration which it prescribes."

EDUCATIONAL TIMES.—"A better course of experimental optics than this can hardly be desired. . . . A work which, as a clear and thoughtful treatise on one of the most interesting of modern sciences, we can heartily recommend to all."

MACMILLAN AND CO., LTD., LONDON.

MESSRS. MACMILLAN AND CO.'S BOOKS
FOR
STUDENTS OF PHYSICS.

LESSONS IN ELEMENTARY PHYSICS. By BALFOUR STEWART, LL.D., F.R.S. With a coloured spectrum. New and Enlarged Edition. Fcap. 8vo, 4s. 6d.
QUESTIONS. By T. H. CORE. Pott 8vo, 2s.

LESSONS IN ELEMENTARY PRACTICAL PHYSICS. By BALFOUR STEWART, LL.D., F.R.S., and W. W. HALDANE GEE, Demonstrator and Assistant Lecturer in Physics, the Owens College. Vol. I. General Physical Processes. Crown 8vo, 6s. Vol. II. Electricity and Magnetism. Crown 8vo, 7s. 6d.

PROBLEMS AND QUESTIONS IN PHYSICS. By CHARLES P. MATTHEWS, M.E, Associate Professor of Electrical Engineering, Purdue University, formerly Instructor in Physics, Cornell University, and JOHN SHEARER, B.S., Instructor in Physics, Cornell University. 8vo, 7s. 6d. net.

ELEMENTS OF THEORETICAL PHYSICS. By Dr. C. CHRISTIANSEN, Professor of Physics in the University of Copenhagen. Translated into English by W. F. MAGIE, Ph.D., Professor of Physics in Princeton University. 8vo, 12s. 6d. net.

LECTURES ON SOME RECENT ADVANCES IN PHYSICAL SCIENCE. By P. G. TAIT, M.A., Sec. R.S.E., formerly Fellow of St. Peter's College, Cambridge; Professor of Natural Philosophy in the University of Edinburgh. Third Edition. Crown 8vo, 9s.

HEAT. By Prof. P. G. TAIT. Crown 8vo, 6s.

THE THEORY OF HEAT. By THOMAS PRESTON, M.A. (Dub.), Fellow of the Royal University of Ireland, and Professor of Natural Philosophy, University College, Dublin. 8vo, 17s. net.

THE THEORY OF LIGHT. By T. PRESTON, M.A. Second Edition. 8vo, 15s. net.

BURNETT LECTURES—ON LIGHT. In Three Courses. Delivered at Aberdeen in November 1883, December 1884, and November 1885. By Sir GEORGE GABRIEL STOKES, M.A., F.R.S., Fellow of Pembroke College, and Lucasian Professor of Mathematics in the University of Cambridge. Second Edition. Crown 8vo, 7s. 6d.
[*Nature Series.*]

ELECTRIC WAVES. Being Researches on the Propagation of Electric Action with Finite Velocity through Space. By HEINRICH HERTZ, late Professor of Physics in the University of Bonn. Authorised Translation by D. E. JONES, B.Sc. With Preface by Lord KELVIN, P.R.S. Illustrated. 8vo, 10s. net.

MACMILLAN AND CO., LTD., LONDON.

MESSRS. MACMILLAN AND CO.'S BOOKS
FOR
STUDENTS OF PHYSICS.

MISCELLANEOUS PAPERS. By HEINRICH HERTZ, late Professor of Physics in the University of Bonn. With an Introduction by Prof. PHILIPP LENARD. Authorised Translation by D. E. JONES, B.Sc., and G. A. SCHOTT, B.A., B.Sc. Demy 8vo, 10s. net.

PAPERS ON ELECTROSTATICS AND MAGNETISM. By Lord KELVIN, P.R.S. Second Edition. 8vo, 18s.

MODERN VIEWS OF ELECTRICITY. By OLIVER J. LODGE, D.Sc., LL.D., F.R.S., Professor of Experimental Physics in University College, Liverpool. With Illustrations. Crown 8vo, 6s. 6d. [*Nature Series*.

LABORATORY MANUAL OF PHYSICS AND APPLIED ELECTRICITY. Edited by E. L. NICHOLS. Vol. I. Junior Course in General Physics. By E. MERRITT and F. J. ROGERS. 12s. 6d. net. Vol. II. Senior Course. By G. S. MOLER, F. BEDELL, H. J. HOTCHKISS, C. P. MATTHEWS, and Editor. 8vo, 12s. 6d. net.

THE THEORY AND PRACTICE OF ABSOLUTE MEASUREMENTS IN ELECTRICITY AND MAGNETISM. By ANDREW GRAY, M.A., F.R.S.E., Professor of Physics in the University College of North Wales. In Two Vols. Crown 8vo. Vol. I., 12s. 6d.; Vol. II., in Two Parts, 25s.

ABSOLUTE MEASUREMENTS IN ELECTRICITY AND MAGNETISM. By Professor ANDREW GRAY. Second Edition. Fcap. 8vo, 5s. 6d.

A TEXT-BOOK ON ELECTRO-MAGNETISM AND THE CONSTRUCTION OF DYNAMOS. By DUGALD C. JACKSON, B.S., C.E., Professor of Electrical Engineering, University of Wisconsin. 8vo, 9s. net.

A TREATISE ON BESSEL FUNCTIONS AND THEIR APPLICATIONS TO PHYSICS. By A. GRAY, M.A., F.R.S.E., and G. B. MATHEWS, M.A., Fellow of St. John's College, Cambridge. 8vo, 14s. net.

SIR ISAAC NEWTON'S PRINCIPIA. Reprinted for Lord KELVIN, P.R.S., and HUGH BLACKBURN, M.A. 4to, 31s. 6d.

NEWTON'S PRINCIPIA. Sections I., II., III. With Notes and Illustrations. Also a Collection of Problems principally intended as Examples of Newton's Methods. By PERCIVAL FROST, D.Sc., F.R.S., Fellow and Mathematical Lecturer, King's College, Cambridge. 8vo, 12s.

MACMILLAN AND CO., LTD., LONDON.

MESSRS. MACMILLAN AND CO.'S BOOKS
FOR
STUDENTS OF PHYSICS.

AN ESSAY ON THE GENESIS, CONTENTS AND HISTORY OF NEWTON'S "PRINCIPIA." By WALTER W. ROUSE BALL, Fellow and Assistant Tutor of Trinity College, Cambridge. Crown 8vo, 6s. net.

UTILITY OF QUATERNIONS IN PHYSICS. By ALEXANDER M'AULAY, M.A., Lecturer in Mathematics and Physics in the University of Tasmania. 8vo, 5s. net.

AN ELEMENTARY TREATISE ON THEORETICAL MECHANICS. By Professor ALEXANDER ZIWET, Assistant Professor of Mathematics in the University of Michigan. Part I. Kinematics. Part II. Introduction to Dynamics; Statics. 8vo, 8s. 6d. net each part.

APPLICATIONS OF DYNAMICS TO PHYSICS AND CHEMISTRY. By J. J. THOMSON, M.A., F.R.S., Fellow of Trinity College, and Cavendish Professor of Experimental Physics, Cambridge. Crown 8vo, 7s. 6d.

A TREATISE ON THE MOTION OF VORTEX RINGS. An Essay to which the Adams Prize was adjudged in 1882, in the University of Cambridge. By J. J. THOMSON, F.R.S. 8vo, 6s.

THE ELEMENTS OF GRAPHIC STATICS. A Text-Book for Students of Engineering. By L. M. HOSKINS, Professor of Pure and Applied Mathematics in the Leland Stanford Junior University, formerly Professor of Mechanics in the University of Wisconsin. 8vo, 10s. net.

A TREATISE ON HYDROSTATICS. By ALFRED GEORGE GREENHILL, F.R.S., Professor of Mathematics in the Artillery College, Woolwich. Crown 8vo, 7s. 6d.

A TEXT-BOOK OF THE PRINCIPLES OF PHYSICS. By ALFRED DANIELL, M.A., Lecturer on Physics in the School of Medicine, Edinburgh. Third Edition. Medium 8vo, 21s.

PHYSICS FOR STUDENTS OF MEDICINE. By ALFRED DANIELL, M.A., LL.B., D.Sc., F.R.S.E., Examiner in Physics to the Royal College of Physicians of Edinburgh. Fcap. 8vo, 4s. 6d.

PHYSICS. Advanced Course. By GEORGE F. BARKER, Professor of Physics in the University of Pennsylvania. With a Magnetic Map of the United States. 8vo, 21s.

MACMILLAN AND CO. LTD., LONDON.

www.ingramcontent.com/pod-product-compliance
Lightning Source LLC
Chambersburg PA
CBHW031439160426
43195CB00010BB/782